TEN THOUSAND
YEARS DEEP

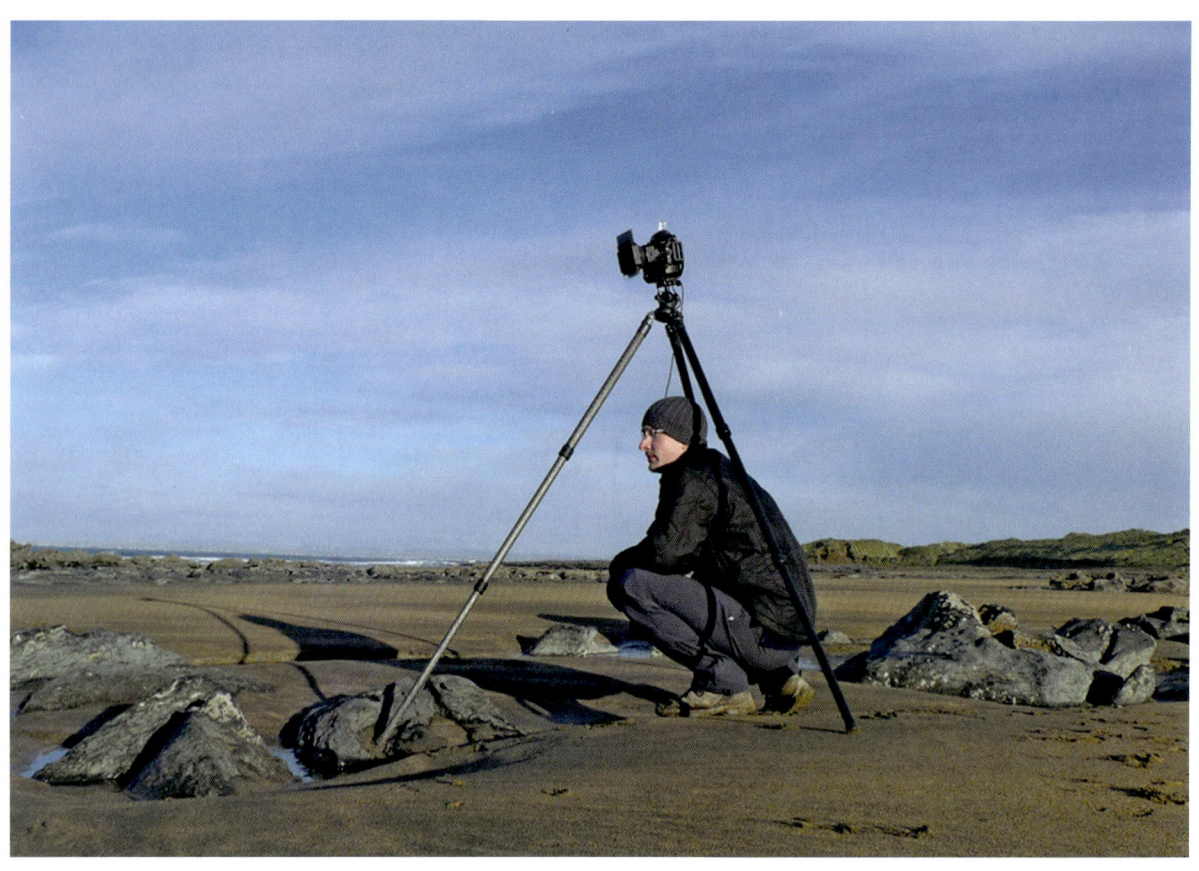

CARSTEN KRIEGER is a photographer, author and environmentalist based on the west coast of Ireland. He has published numerous books on Ireland's landscape, nature and heritage including *Ireland's Islands*, *Ireland's Coast*, *The River Shannon* and *Ireland's Wild Atlantic Way*. His book *Wild Ireland: A Nature Journey from Shore to Peak* was shortlisted for the An Post Irish Book Awards 2023. His most recent book is *Ireland's Wild Beauty: A Book of Days*.

TEN THOUSAND YEARS DEEP

The Story of Ireland's Peatlands

CARSTEN KRIEGER

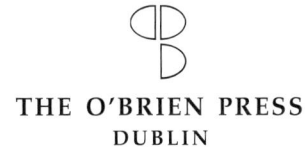

THE O'BRIEN PRESS
DUBLIN

First published 2025 by The O'Brien Press Ltd,
12 Terenure Road East, Rathgar, Dublin 6, D06 HD27, Ireland.
Tel: +353 1 4923333; Fax: +353 1 4922777
E-mail: books@obrien.ie; Website: obrien.ie
The O'Brien Press is a member of Publishing Ireland.

ISBN: 978-1-78849-486-1

Text & photography © copyright Carsten Krieger 2025
The moral rights of the author have been asserted.
Copyright for typesetting, layout, editing, design © The O'Brien Press Ltd
Design, layout and cover design by Tanya M Ross, www.elementinc.ie

Quote on p7 from *Stirring the Mud: On Swamps, Bogs, and Human Imagination* by Barbara
Hurd (University of Georgia Press, 2008)

8 7 6 5 4 3 2 1
29 28 27 26 25

Printed by L&C Printing Group, Poland.

The paper in this book is produced using pulp from managed forests

To the best of our knowledge, this book complies in full with the requirements of the
General Product Safety Regulation (GPSR). For further information and help with any
safety queries, please contact us at productsafety@obrien.ie.

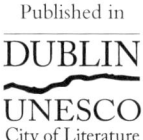

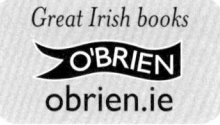

Table of Contents

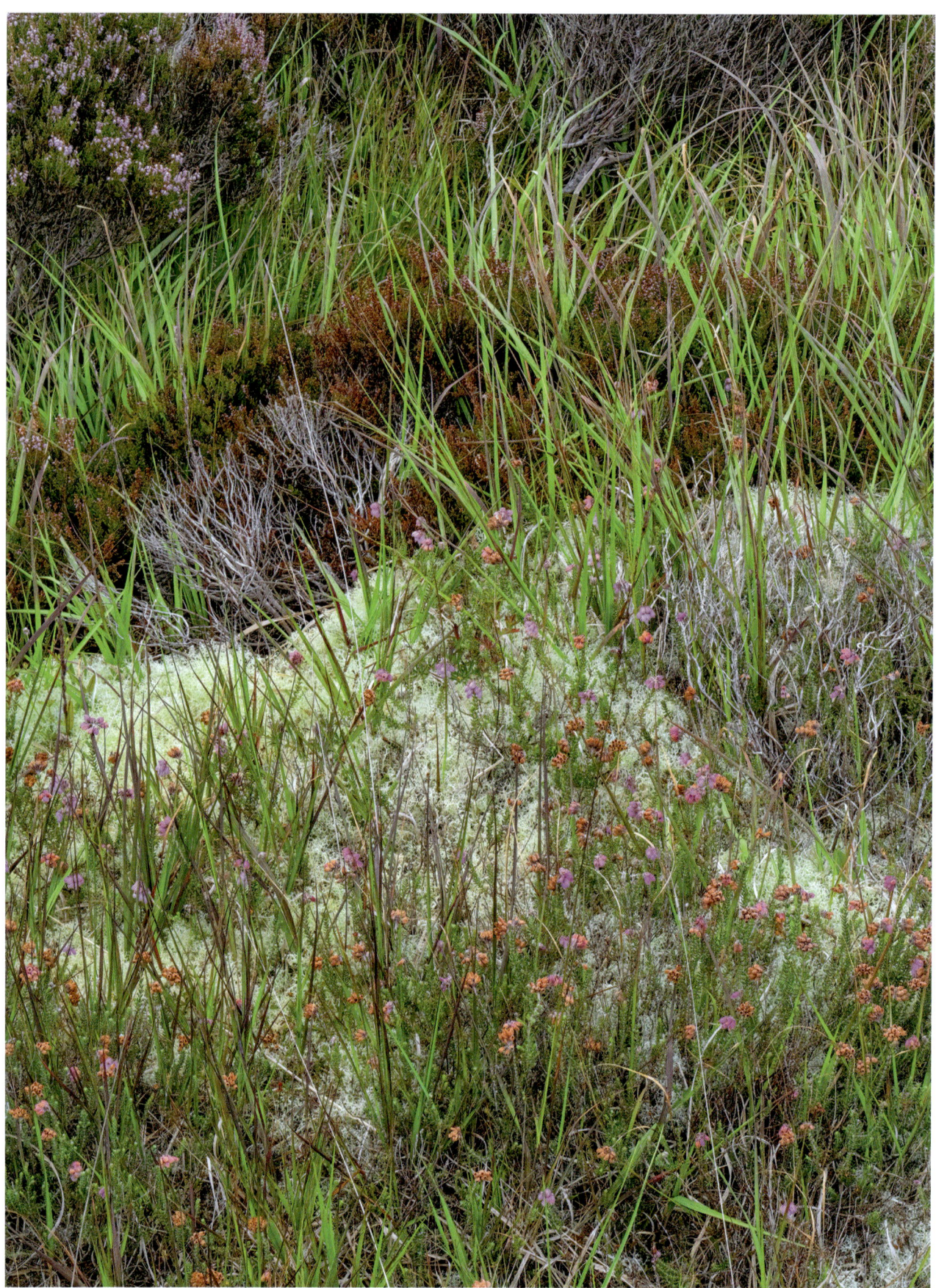

Blanket bog in late summer with heathers, lichen and grasses

Swamps and bogs are places of transition and wild growth, breeding grounds, experimental labs where organisms and ideas have the luxury of being out of the spotlight, where the imagination can mutate and mate, send tendrils into and out of the water.

Barbara Hurd,
from *Stirring the Mud: On Swamps, Bogs, and Human Imagination*

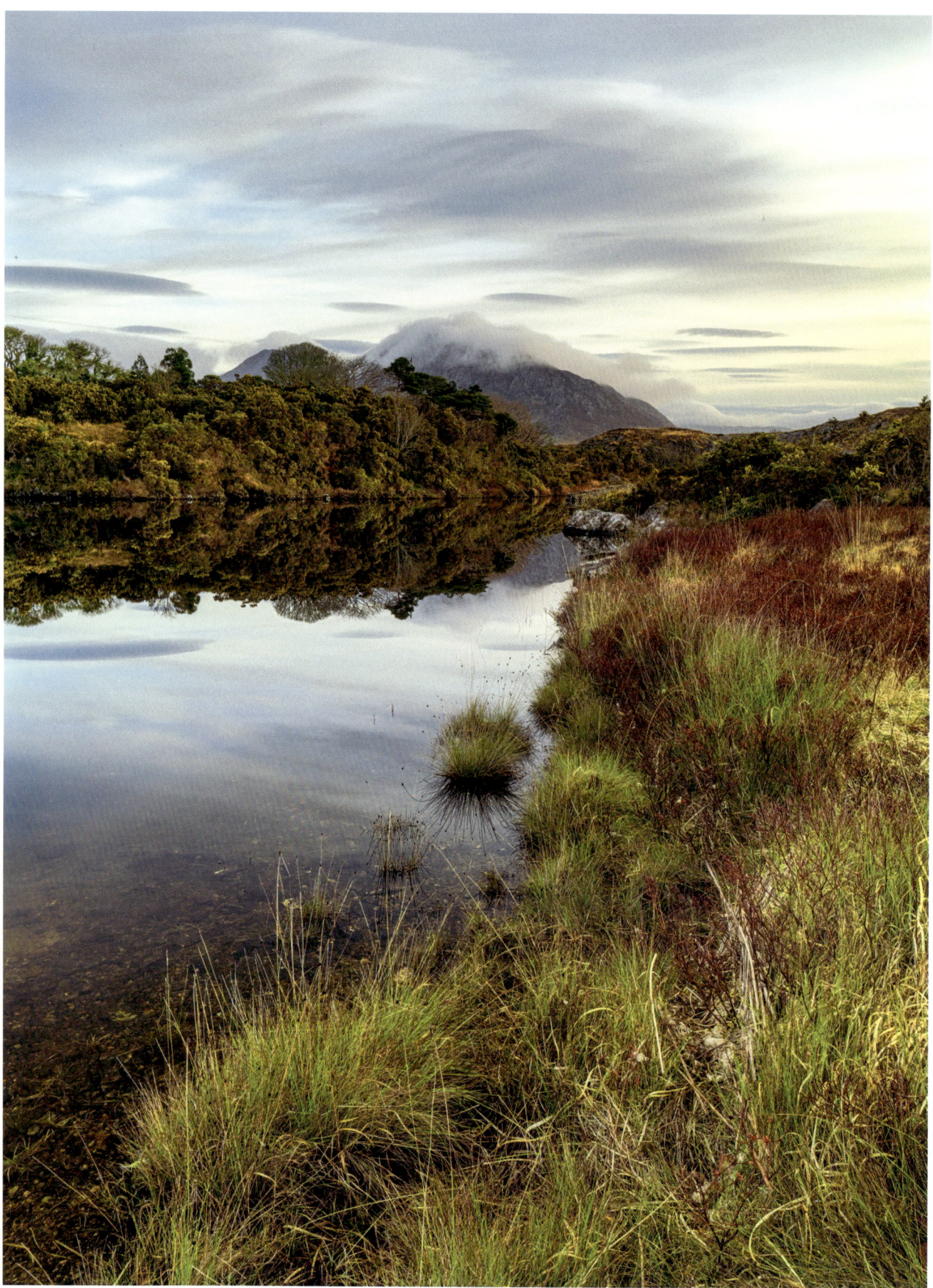

Blanket bog in Connemara

INTRODUCTION

They cover more than one million hectares of Ireland's land mass. They can be found at the coast and in the mountains. They have helped to shape the country's history and heritage, and they are the most intriguing habitat there is – for me, at least. They are our peatlands.

I spent my first year in Ireland in an old cottage in the middle of Shragh Bog near Doonbeg in County Clare. This bog is typical for Ireland, partly cultivated for farming and partly used as a domestic fuel source. While it is far from its original state, I couldn't escape its lure. A few steps out the back door would bring me onto the bog road, a pothole-covered track giving access to the turf banks, running past dark Sitka spruce plantations, along pastures divided by rows of the massive root structures of bog pines, dug out from the numerous turf banks. Here and there, pockets of living and growing bog stood out, a symphony of the vibrant green, red and yellow of sphagnum mosses, dotted with tiny red sundews, green sedges and grey lichen.

In spring and summer, the skylark would sing high above, interrupted from time to time by the call of the cuckoo and the ghostly drumming of the snipe. In April and May, turf sods would appear, laid out neatly to dry on top of the turf banks and beside the road. Later, these sods would be footed into stacks and the common cotton grass, better known as bog cotton, would grow around them, creating a sea of fluffy, snow-white heads bopping in the breeze. In places, heathers, marsh orchids and bog asphodel would thrive, and dragonflies and damselflies would show off their air acrobatics over the pools that had formed beside the turf banks. In late summer, tractors and trailers would turn up to bring home the turf, and in autumn, the rich green of summer would be replaced by a variety of warm brown tones, glowing in the light of the rising and setting sun. Then the bog would turn eerily quiet. Heavy rain would leave the remains of grasses, sedges and heathers covered in a glistening film of water, which would turn into a sparkling frost during the coldest of the winter nights.

Ever since these days, more than twenty years ago, I have tried to learn as much as I can about Ireland's peatlands; to explore them, and to understand them. This book is the result. I hope you'll enjoy the journey.

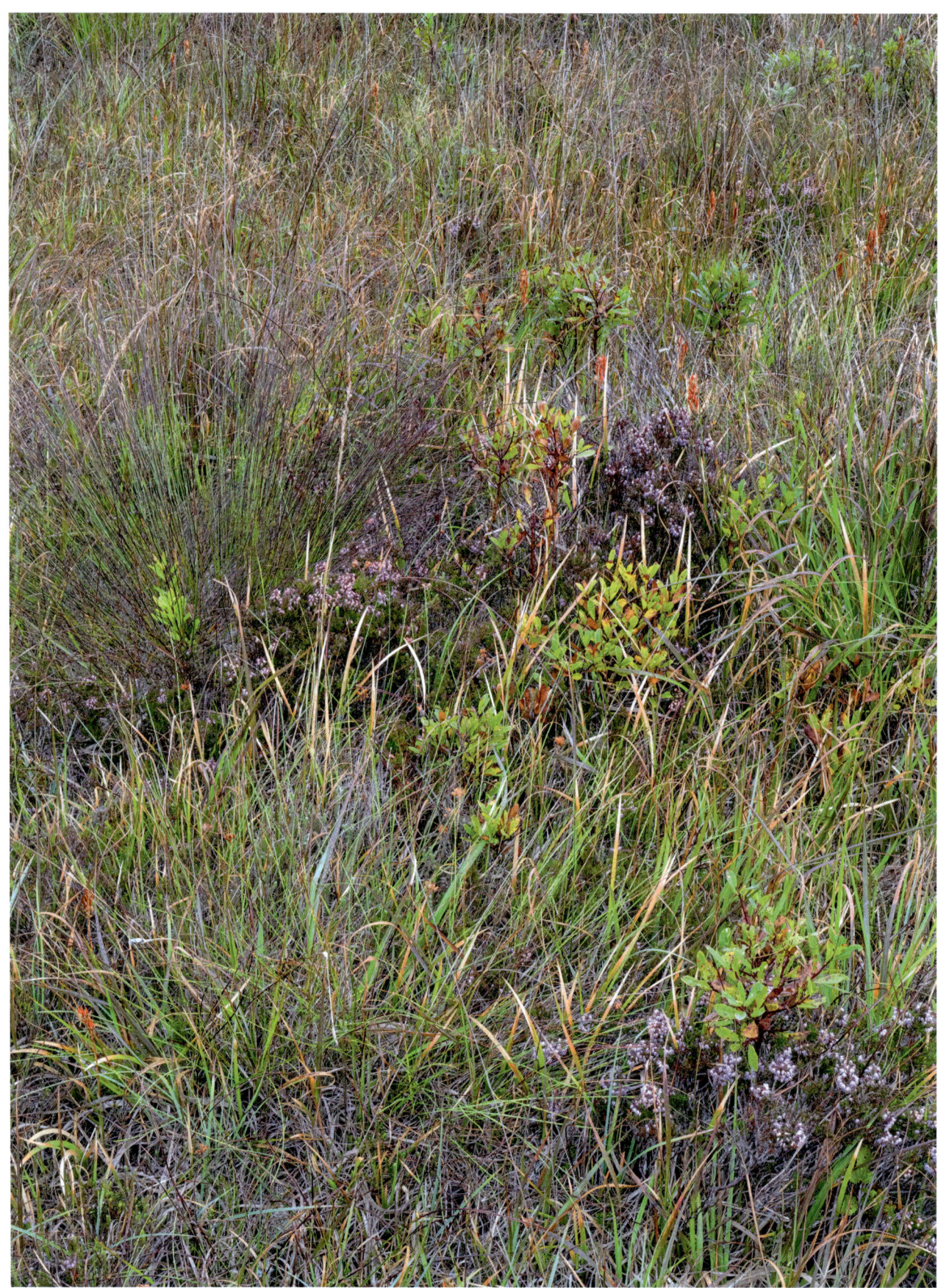

Autumnal blanket bog with heather, bog myrtle and grasses

Chapter One
INTO THE PEATLANDS

What are peatlands? It sounds like a simple question, but the answer is anything but straightforward. Peatlands are as diverse and complex as they are sublime and enchanting. In summer, they appear in the vibrant green of grasses and sedges, dotted with the white, yellow, red, pink, blue and purple of a plethora of flowering plants. In winter, the same landscape reverts to a collection of warm yellow and brown tones. Their range extends from the tropics to subarctic regions and from sea level into the mountains. Worldwide, peatlands cover an estimated 500 million hectares, or 4% of the planet's land area. Peatlands can appear as vast plains where grasses, mosses, lichen and other plants blend together to create thick, entangled carpets. They can feature dense thickets of shrub and small woodlands that echo with the sound of birdsong. Pools, ponds and lakes pop up here and there, sometimes surrounded by large reed beds and sometimes treacherously invisible, concealed under a carpet of shimmering sphagnum mosses.

It is this diversity and complexity that makes peatlands so hard to define. These landscapes have been labelled – among other names – as bog, quagmire, marsh, swamp, mire, morass, swale, moss, heath, fen, palsa, flark and lagg. All of these words originated in everyday language, and in recent decades some have been adopted into scientific circles. However, because their meanings have not yet been widely and clearly defined, the same word can mean something different to different people and in different places.

At first sight, it's hard to distinguish between the different types of peatland. For example, there is no apparent difference between a wet heath and a blanket bog, and a fen can look very much like a raised bog in places. Only a closer look reveals the variations in the habitats' ecology. To make the situation even more complicated, peatland habitats tend to mingle, so it is difficult if not impossible to say where one ends and the other begins. Some of these habitats also change over time, and what starts out as a fen can slowly transform into a raised bog.

"

DEAD PLANTS AND ANIMALS REMAIN IN LIMBO, NEITHER FULLY PRESERVED NOR FULLY DECOMPOSED

"

There is, of course, one feature they all have in common: peat. This unmistakable squelchy substance is made of (mostly) plant and (to a lesser extent) animal remains that have accumulated under (more or less) water-saturated conditions. Oxygen availability in water is considerably lower than it is in air, and the little oxygen available in water-logged ground is quickly used up by microorganisms, resulting in a very low oxygen (hypoxic) or even oxygen free (anoxic) environment. These conditions prevent the full decomposition of any natural matter, so instead of being transformed into soil, dead plants and animals remain in limbo, neither fully preserved nor fully decomposed, and build up over decades, centuries and millennia, layer upon layer. The result of this process is peat, a wet concoction that comes in various shades of russet, its consistency dictated by the plants and animals that compose it. Anything that falls onto the surface or forms part of the top layer of vegetation – leaves, flowers, stems, roots and rhizomes, mosses, lichen, any unfortunate insect, bird or other animal or indeed human – is partially recycled and partially preserved in the growing layers of peat.

Peat has been described and classified in many ways. One is based on its content: sphagnum peat, sedge peat, brown moss peat, woody peat (when small, humified pieces of wood make up more than half of the solid content), lignid peat (when large pieces of tree wood are embedded in the peat) and nano lignid peat (when remains of shrubs are visible in the peat). Another method looks at the decomposition stage and differentiates between fibric or white peat (the top and least decomposed layer, where plant fibres are still clearly visible), hemic or grey peat (the partially humified middle layer) and sapric or black peat (the fully humified bottom layer). The latter is recognised as the most valuable peat for heating and cooking, renowned for its hot flame and long burning time. Before peat inspired any scientific interest, it was only classified by its value as a fuel. In his 17th-century book *A Natural History of Ireland*, physician Gerard Boate described two sorts: 'That which is taken out of the dry-bogs, or red-bogs, is light, spungy, of a reddish colour, kindleth easily, and burneth very clear, but doth not last. The other on the contrary, which is raised out of the green-bogs or wet bogs, is heavy, firm, black, doth not burn soon, nor with so great a flame, but lasteth a long while, and maketh a very hot fire, and leaveth foul yellowish ashes.'

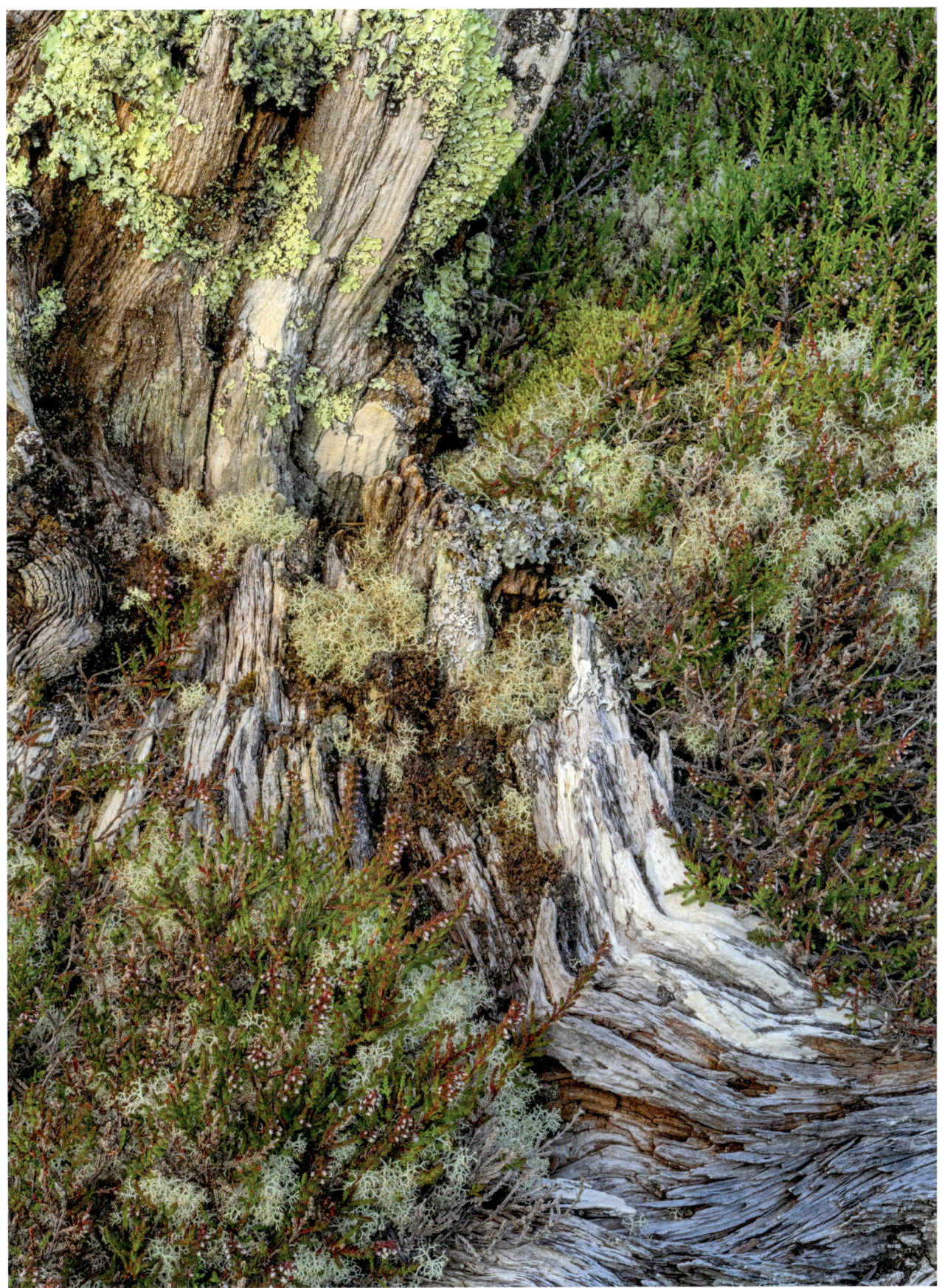

Bog pine with heathers and lichens

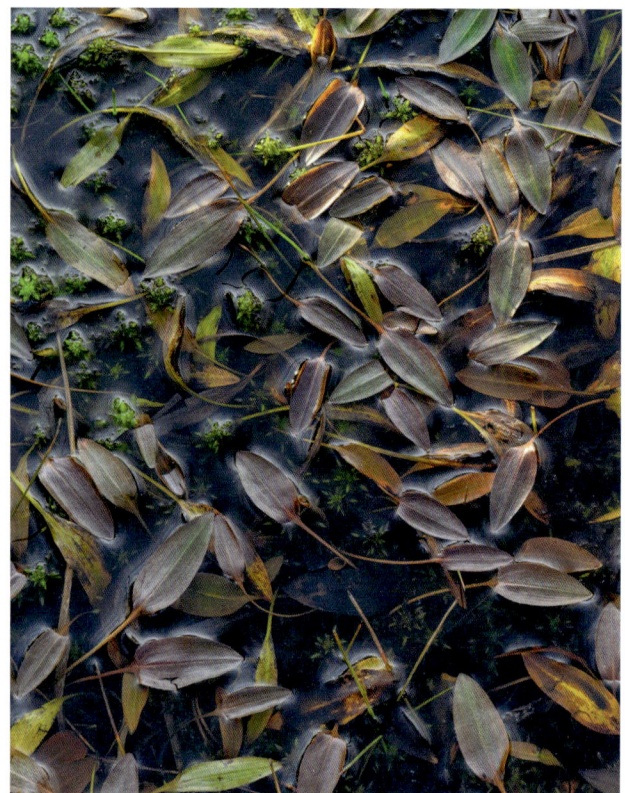

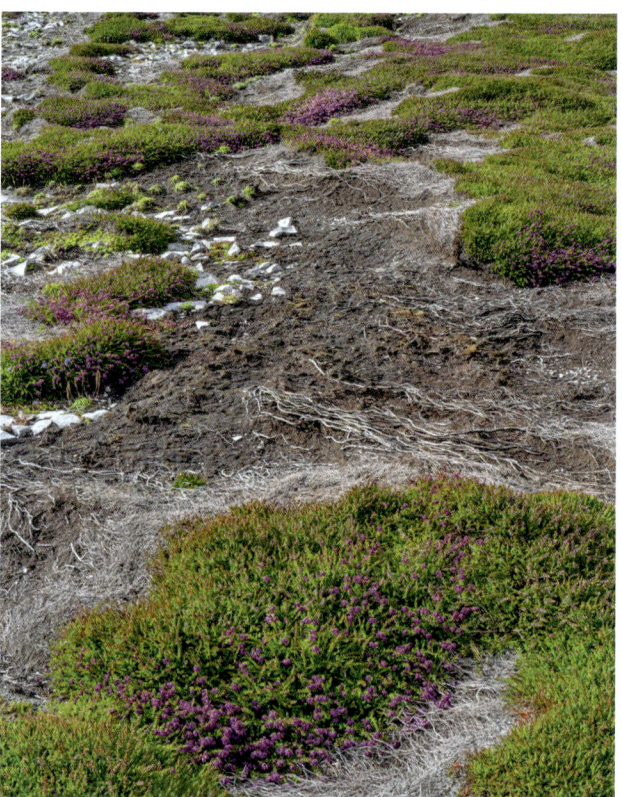

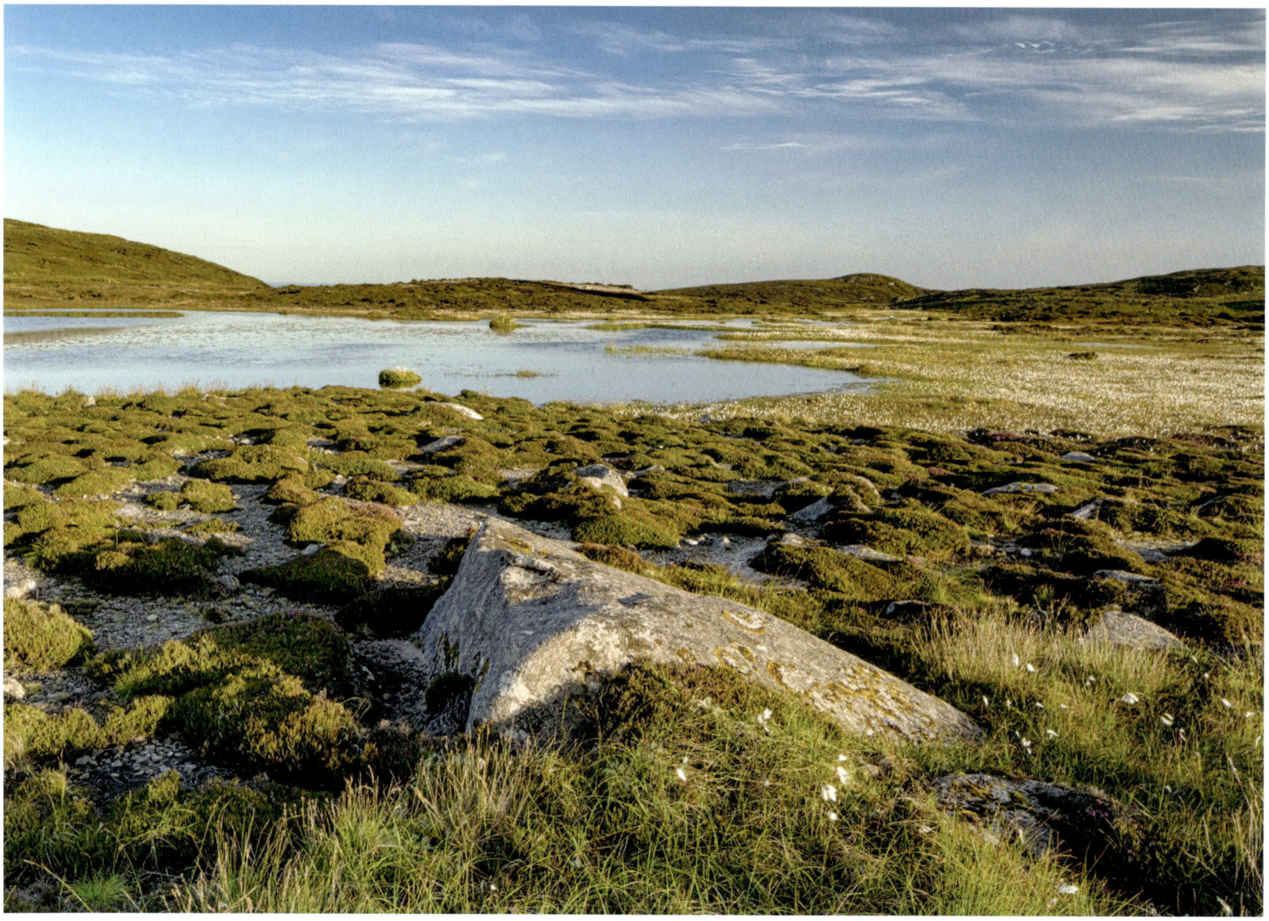

Opposite top left: Bog pondweed
Opposite top right: Montane heath, Loop Head, County Clare
Opposite bottom: Blanket bog, Arranmore Island, County Donegal

A peatland becomes a peatland, in the scientific sense, when the overall peat layer is thicker than thirty centimetres (according to the International Mire Conservation Group, 2002). For this book, however, the net will be cast a bit wider and include habitats like wet woodlands, heathlands and salt-marshes that also have peat layer, albeit a thin one.

In Ireland, peat has a growth rate of around one millimetre per year. Since their formation, the blanket bogs of Ireland's west coast have produced peat layers accumulating to up to five metres; in the raised bogs of the midlands, the peat has grown to a thickness of twelve metres and would have grown even more if men hadn't discovered it as a fuel source. The Philippi Moor in Greece – which escaped the glaciation of the Ice Age and has been mostly left alone ever since its formation – boasts peat layers up to 190 metres thick, formed over a period of 1.35 million years.

Each layer of peat is a page in an immense history book going back to the time the very first layer was formed. In Ireland, this was the end of the last glaciation some 10,000 years ago.

Smaller plants fully transform into peat, with only their pollens and spores remaining identifiable. These pollens and spores, perfectly preserved in the peat, have allowed researchers to piece together the natural history of Ireland from the tundra stage that followed the retreat of the ice sheets; through the arrival of the first trees, namely birch and hazel; the subsequent rise of the vast oak and pine forests; and their demise at the hands of humans and replacement with grass- and farmland. Bigger plants, meanwhile, keep some of their shape and are fully or partly preserved inside the peat; we can find bog oaks and bog pines, branches, trunks and root structures of trees that were engulfed by the growing peat. Animal remains like those of the Irish elk have also been found in bogs. Together, they become the peat archive, a visual history book that takes the reader back to the end of the Ice Age.

Also embedded in the peat are chapters of human history. Artefacts like gold and silver jewellery, amber beads, bronze and flint spear-heads, stone axe heads, leather shoes and preserved butter – aptly known as bog butter – have been found in raised and blanket bogs all over Ireland. Entire settlements like the 7,000-year-old Meso-lithic habitations at Boora in County Offaly or the 5,000-year-old Neolithic farming landscape of the Céide Fields in County Mayo have been discovered under the peat layers. Dating to medieval times are toghers, causeways built across the bog from either woven hurdles resting on tightly packed brushwood, which were likely only used for foot traffic, or more sophisticated roads constructed from wooden planks placed on rails, which would have been suitable for horses and carts. One of the best preserved toghers is the Corlea Trackway, also known as Danes' Road, in County Longford. A total of five different tracks have been found here. The oldest, made of brushwood, dates to 1020 BC. A more refined and wider bog road, made from oak planks supported by birch pegs, is thought to have been built around 150 BC. Excavations in Lemanaghan Bog in County Offaly also revealed tracks that show the evolution of the toghers. Here, several roads seem to have been built on top of

"

SINCE THE LATE 18TH CENTURY, MORE THAN A HUNDRED HUMAN BODIES OR PARTS THEREOF WERE UNCOVERED IN BOGS

"

each other, the oldest also made of brushwood, the most recent of flagstone. Because these trackways seemingly covered only short distances – in the case of the Corlea Trackway, just about one kilometre – and would have lasted at most a few decades before becoming overgrown by the bog or submerged under their own weight, their function is not quite clear. One theory is that toghers might have led to ritual sites in the bog or spots used for peat harvesting. Another, supported by the multi-layered track at Lemanaghan Bog, suggests that at least some of these trackways were more permanent, leading from main roads like the Esker Riada – at the time the main connection from east to west, which allowed safe travelling from Dublin to Galway along a series of ridges rising above the vast peatlands – to small settlements, farming communities and monasteries along the way.

The most telling discoveries made in peatlands might, however, be the bog bodies. Since the late 18th century, more than a hundred human bodies or parts thereof were uncovered in bogs from County Mayo in the west to County Meath in the east. The oldest of these bodies is a 5,000-year-old skeleton from Stoneyisland Bog in County Galway dating to the Neolithic period, and a mummified individual found in Cashel Bog in County Laois.

Stoneyisland Man was discovered at a depth of three metres and is thought to have drowned in the bog some 6,000 years ago. Cashel Man died in the early Bronze Age, around 4,000 years ago, of not-so-natural causes.

There was much speculation about the reasons for these bodies ending up in the peat. Some, like Stoneyisland Man, would likely have died an accidental death; individuals getting lost in the hostile landscape and eventually dying of exposure. Stories supporting this theory are plentiful in local folklore, like the legend of Michael who got caught out by a storm while walking home through Roundstone Bog in Connemara. He sought shelter in a hollow under a large boulder, hoping to sit out the high winds, hail and rain. Michael's body was found many days later in the very same spot, which is still known as Michael's Grave. Other bog bodies might have been the victims of murder or robbery. Some could have been members of the community who had not conformed to societal norms or who had broken the law and so could not be laid to rest alongside their neighbours. It is also possible that people were buried in the bog during times of disease to prevent the spread of the contagion, or for any kind of religious and superstitious beliefs.

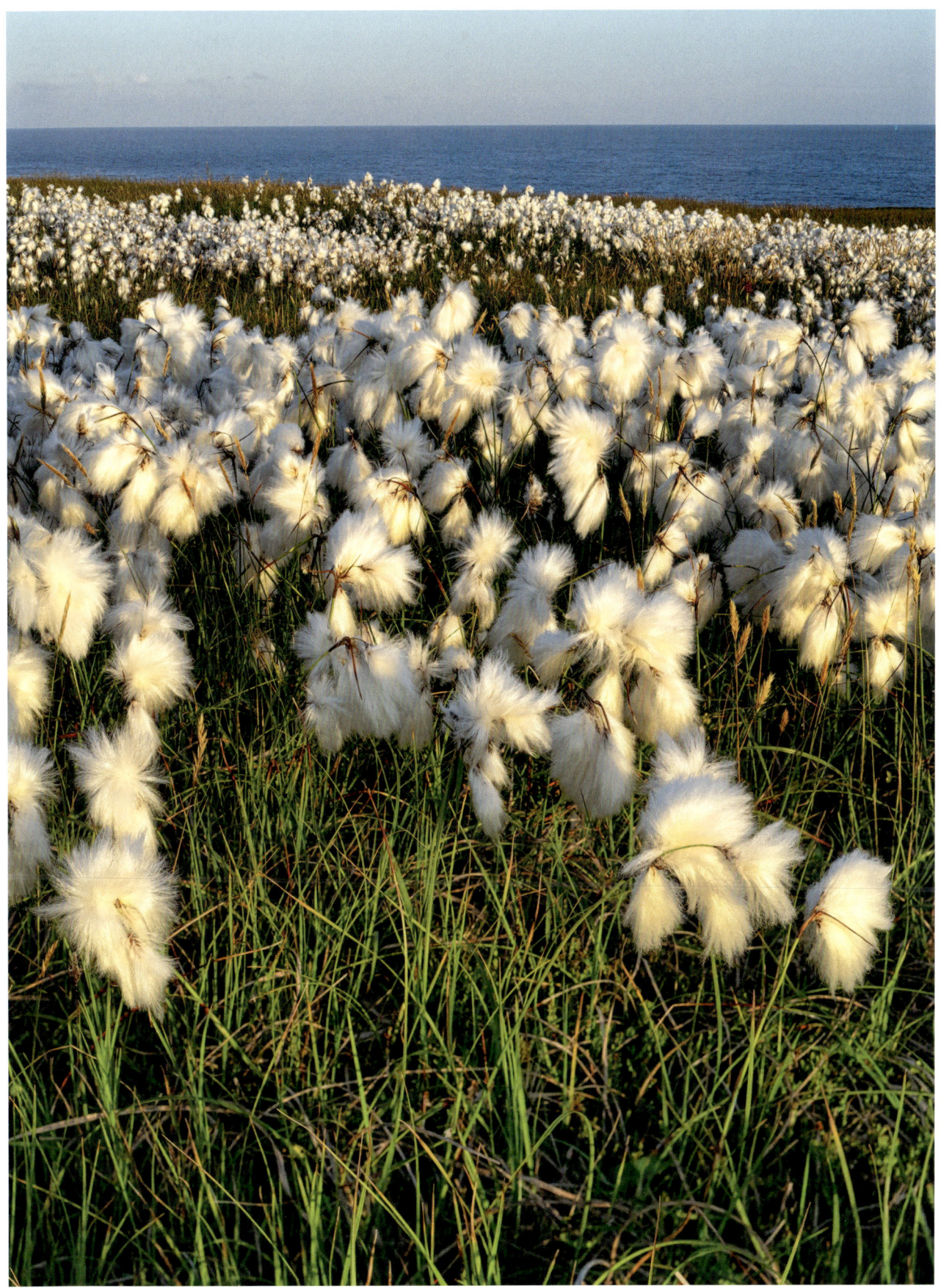

Common cotton grass

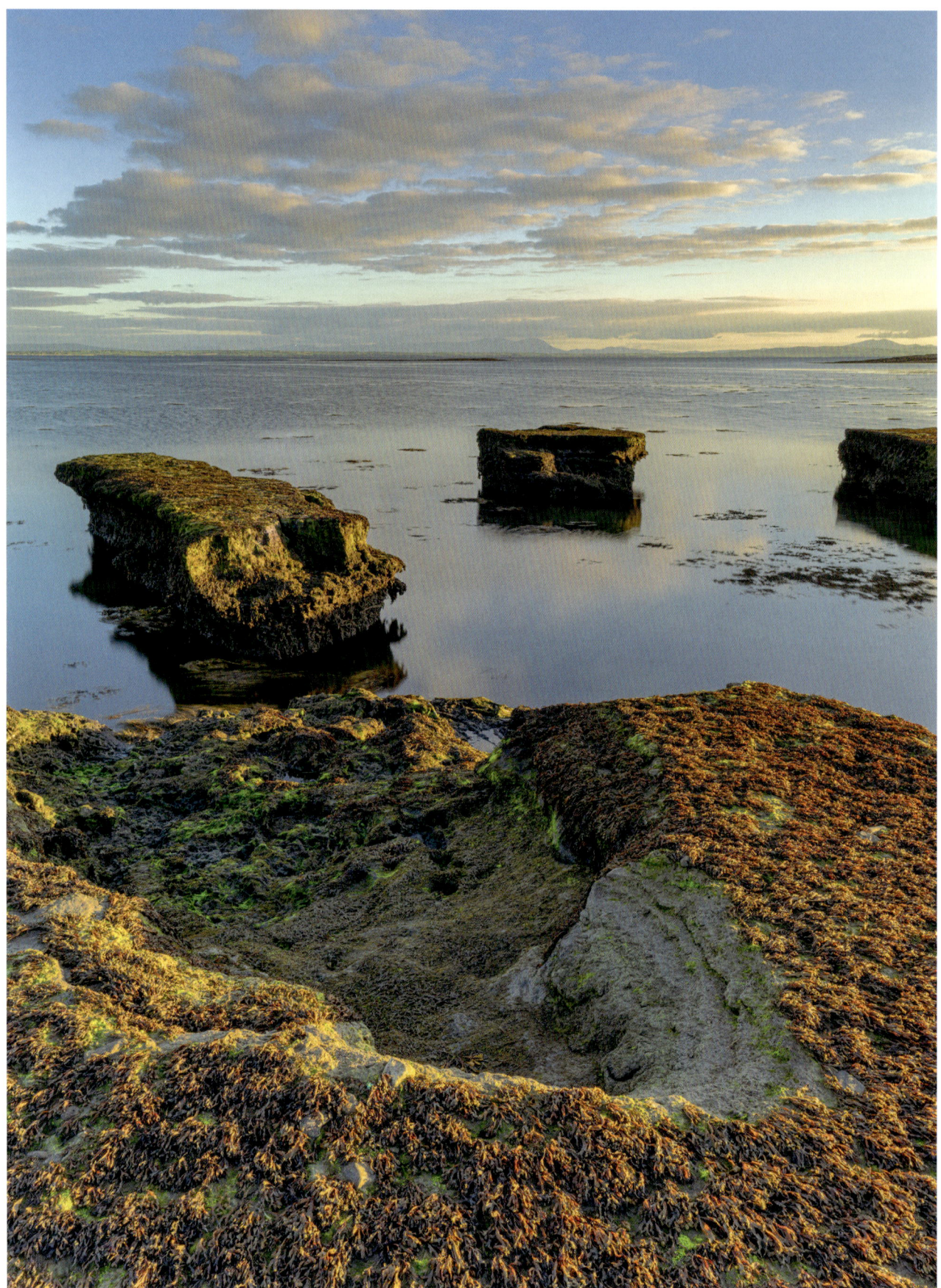

The Drowned Forest, County Clare

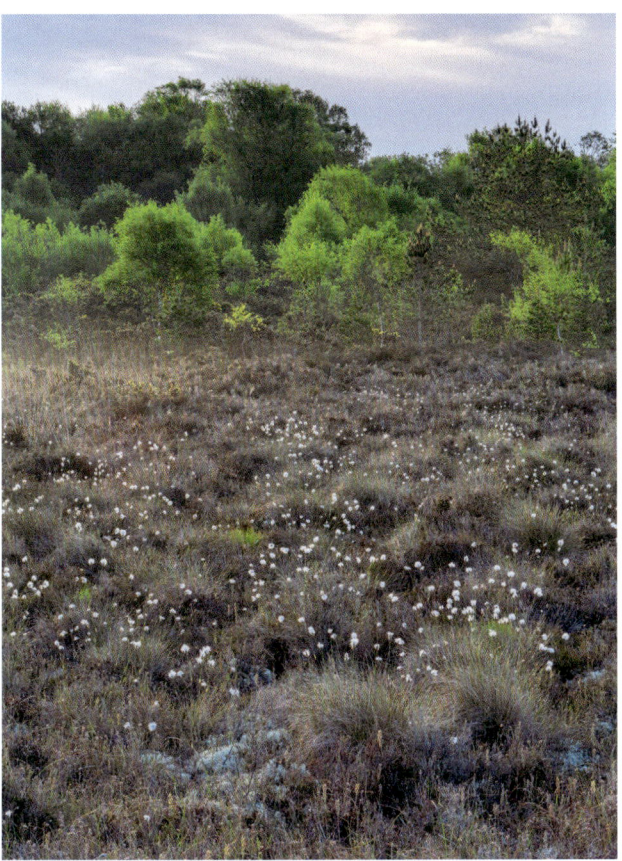

Killaun Bog, County Offaly

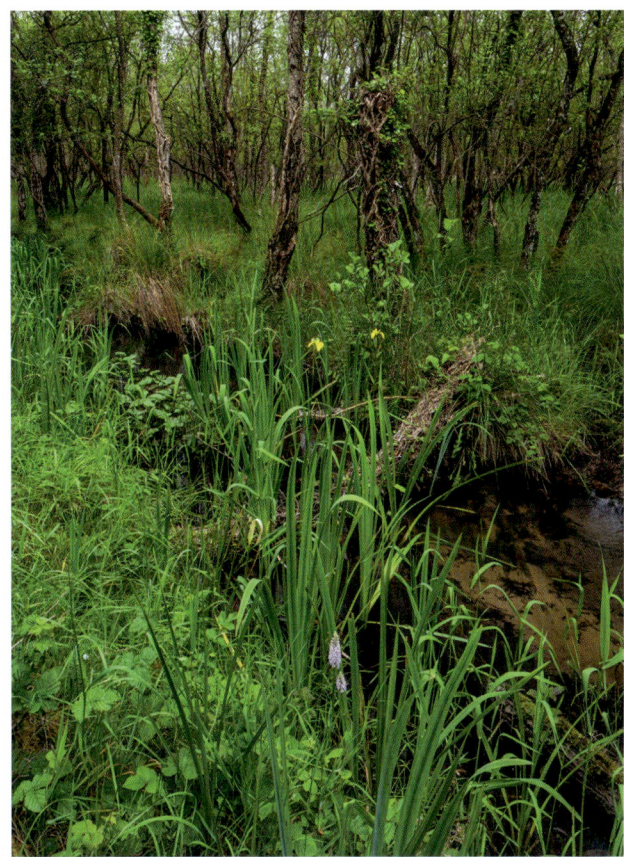

Wet woodland with common spotted orchid and flag iris at Abbeyleix Bog, County Laois

Some of the bodies discovered, however, show signs of a violent death – often a combination of stabbing, hanging and drowning – and a ceremonial burial, with grave markers highlighting the site, and artifacts laid to rest with the deceased. Cashel Man, for example, had suffered a long cut on the lower back and a broken arm, and his resting place was marked with two hazel rods. It is thought that these killings, which started in the early Bronze Age and continued well into the Iron Age, were part of the rituals surrounding the succession of kingship. Unique to these bodies, often discovered at the boundary lines of ancient kingdoms, are signs of what is interpreted as ritualistic torture and sacrifice. Many had their nipples removed, which would have excluded them from becoming king. The Old Croghan Man, discovered in County Offaly in 2003, had a hole in both his upper arms where rope had been fed through to restrain him. Despite the obvious signs of torture, these bodies also carried evidence of high social standing. Old Croghan Man was found with perfectly manicured fingernails, and the nails' nitrogen levels point to a protein-rich, meat-based diet, which was back then reserved for the rich and powerful.

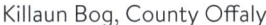

"

IN THE IRISH LANGUAGE, *BOG* MEANS 'SOFT', AND THE WORD WAS ADOPTED INTO ENGLISH IN THE EARLY 16ᵀᴴ CENTURY

"

The peat that became the final resting place of the bog bodies was created by three main factors: water, its potential of hydrogen (better known as pH), and its nutrient content. The finer details of these factors then informed what kind of peatland would develop: raised bog, blanket bog, heath or fen.

Water can either come from above (rain) or below (groundwater). It saturates the ground to various degrees, which is reflected in a water table that is either above, at, or just under the surface. The second factor, the chemical regime defined by the pH level and calcium content of the environment, can either be acidic (the average pH of bogs is 4 and can go as low as 3.5) or base-rich (rich fens have a pH of 7 and above, while poor fens can go as low as 4.5). The third factor is the availability of nutrients, primarily nitrogen (which is needed to build nucleic acids, the building blocks of all life), with phosphorus and potassium playing a smaller role.

The two latter factors, the chemical regime and availability of nutrients, are predominantly shaped by the first, the water source. Groundwater, which can come from springs, flushes, seepages, rivers and lakes, is generally nutrient rich with a high pH and creates the so-called minerotrophic peatlands, the fens. Rainwater, in comparison, is very poor in nutrients and has a low pH, creating slightly acidic habitats, known as ombrotrophic peatlands: the heaths and bogs.

In the Irish language, *bog* means 'soft', and the word was adopted into English in the early 16ᵗʰ century, joining the old English terms 'moor' and 'mire'. 'Mire' made its way into scientific circles after the English botanist Harry Godwin, who helped develop the science of ecology and coined the phrase 'peat archive', suggested using the term to collectively describe blanket bogs, raised bogs, heaths and fens. Today the term 'mire' is only used to describe intact peatlands. Once peatlands have been harvested or drained for agriculture, they are scientifically no longer classified as a mire, because the peat-forming plants on the surface have been removed to get to the peat below or replaced by crops. In that context, 'mire' is mostly used in a botanical and ecological context to describe a wet terrain dominated by the most characteristic peat-forming plants, which are sphagnum mosses, grasses and sedges. The origins of the word go back to Old Norse, where *myrr* described a boggy, swampy ground. The related term 'moor', which conventionally describes any tract of uncultivated land covered in heather, grasses, sedges or similar vegetation, can also be traced back to Old Norse where it appeared as *mörr*, as well as other European languages like the Dutch *meer*, the Old High German *muor* and the Old English *mor*.

While in literature and everyday language, 'moor' is often used interchangeably with 'heath' – 'moor' mainly being used to describe upland areas and 'heath' being attributed to the lowlands – scientifically 'heath' describes a distinct group of peatland habitats characterised by a very thin peat layer and typical flora. Occupying a bit of an ecological grey area are swamps and marshes, which are first and foremost classified as wetlands.

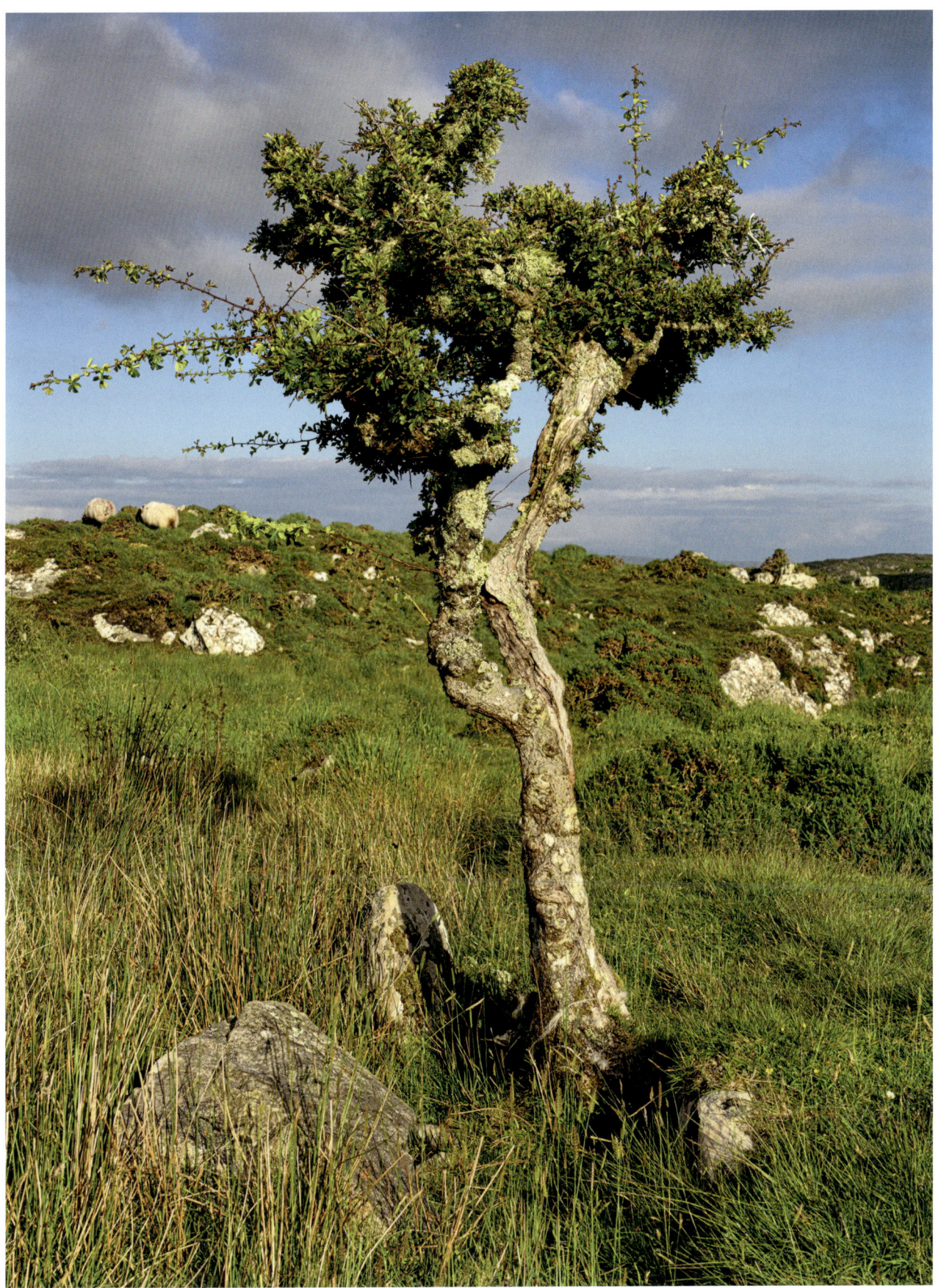

Hawthorn, Roundstone Bog, Connemara

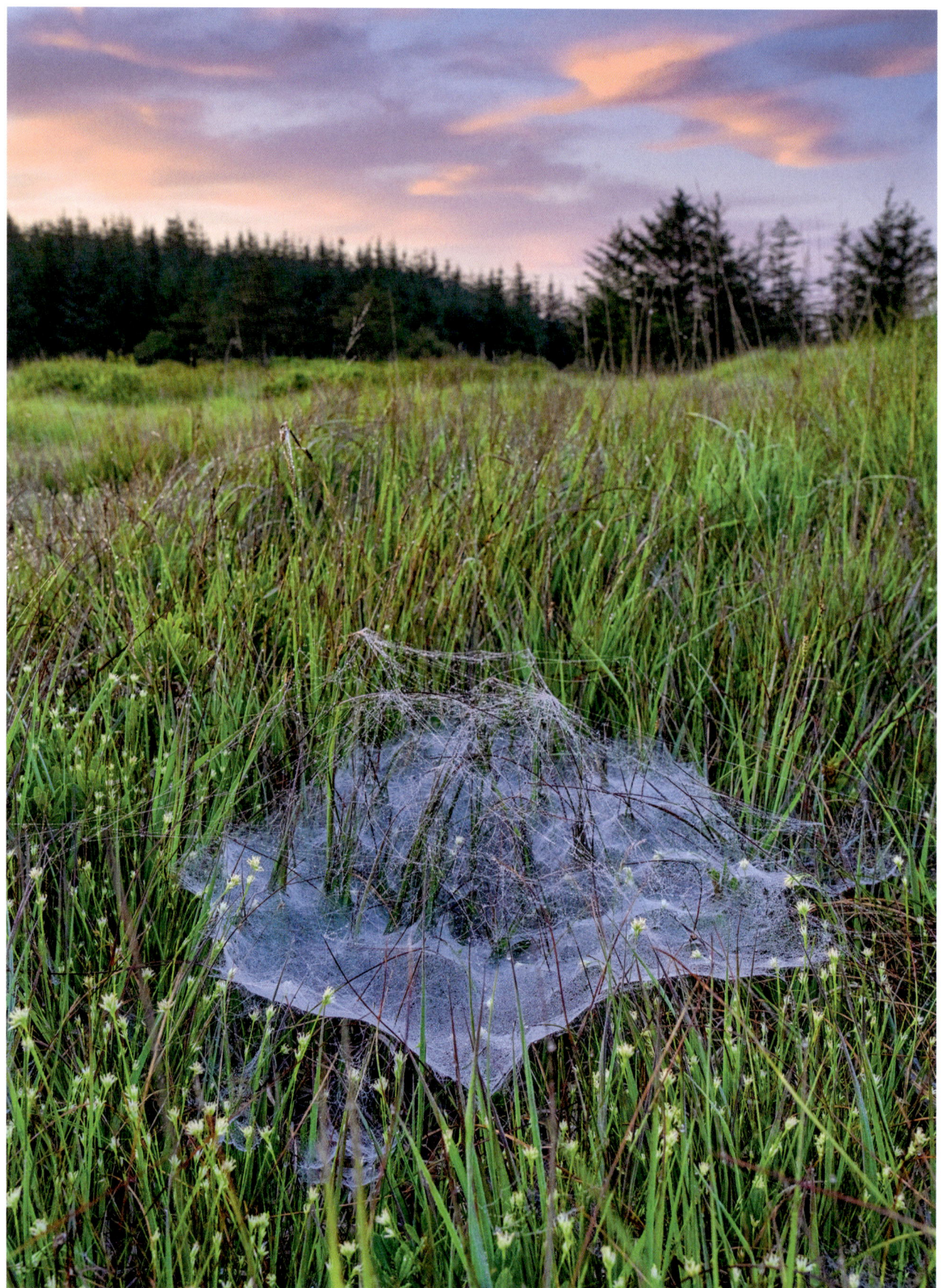

Cutover bog, Shragh Bog, County Clare

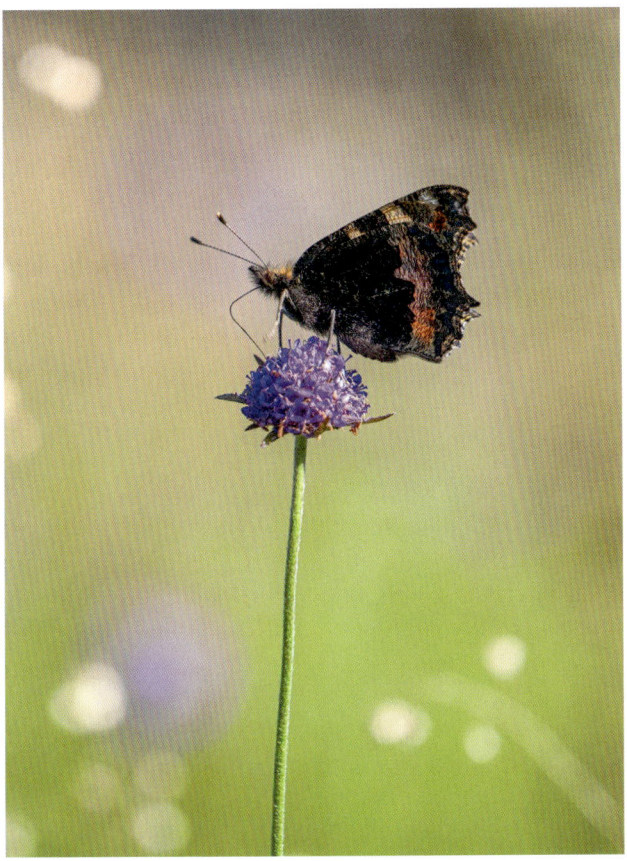

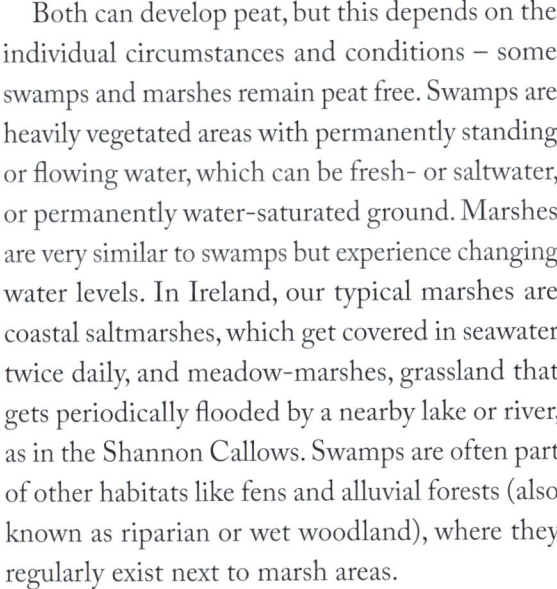

Small tortoiseshell on devil's bit scabious

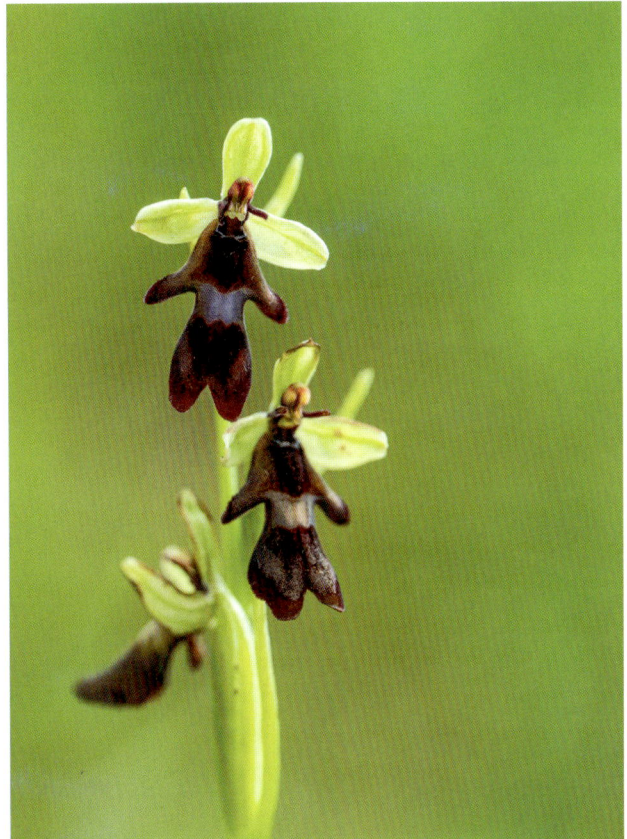

Fly orchid

Both can develop peat, but this depends on the individual circumstances and conditions – some swamps and marshes remain peat free. Swamps are heavily vegetated areas with permanently standing or flowing water, which can be fresh- or saltwater, or permanently water-saturated ground. Marshes are very similar to swamps but experience changing water levels. In Ireland, our typical marshes are coastal saltmarshes, which get covered in seawater twice daily, and meadow-marshes, grassland that gets periodically flooded by a nearby lake or river, as in the Shannon Callows. Swamps are often part of other habitats like fens and alluvial forests (also known as riparian or wet woodland), where they regularly exist next to marsh areas.

Ireland holds more than one million hectares of peatland, close to 20% of the total land area. This is not only more than any other European country apart from Finland, but it also makes our small island one of the countries with the largest peatland cover globally. The most widespread of the Irish peatlands is blanket bog, accounting for 77% of all peatlands. Subdivided into Atlantic (or lowland) blanket bogs and mountain blanket bogs, they are mainly located in the west of the country, from County Cork in the southwest to County Donegal in the northwest. Raised bogs represent 17% of our peatlands and cluster in the midlands, especially in the counties Offaly, Kildare and Laois. Here, the vast Bog of Allen once covered over 900 square kilometres. These bogs, and the blanket bogs in particular, are fuelled by high amounts of rainfall produced by alternating low pressure systems pushing in from the Atlantic Ocean, while the North Atlantic Drift, better known as the Gulf Stream, keeps temperatures mild all year round, creating the perfect growing conditions for the peatland flora. The remaining 6% of Irish peatlands consist of fens, heaths and other peatlands.

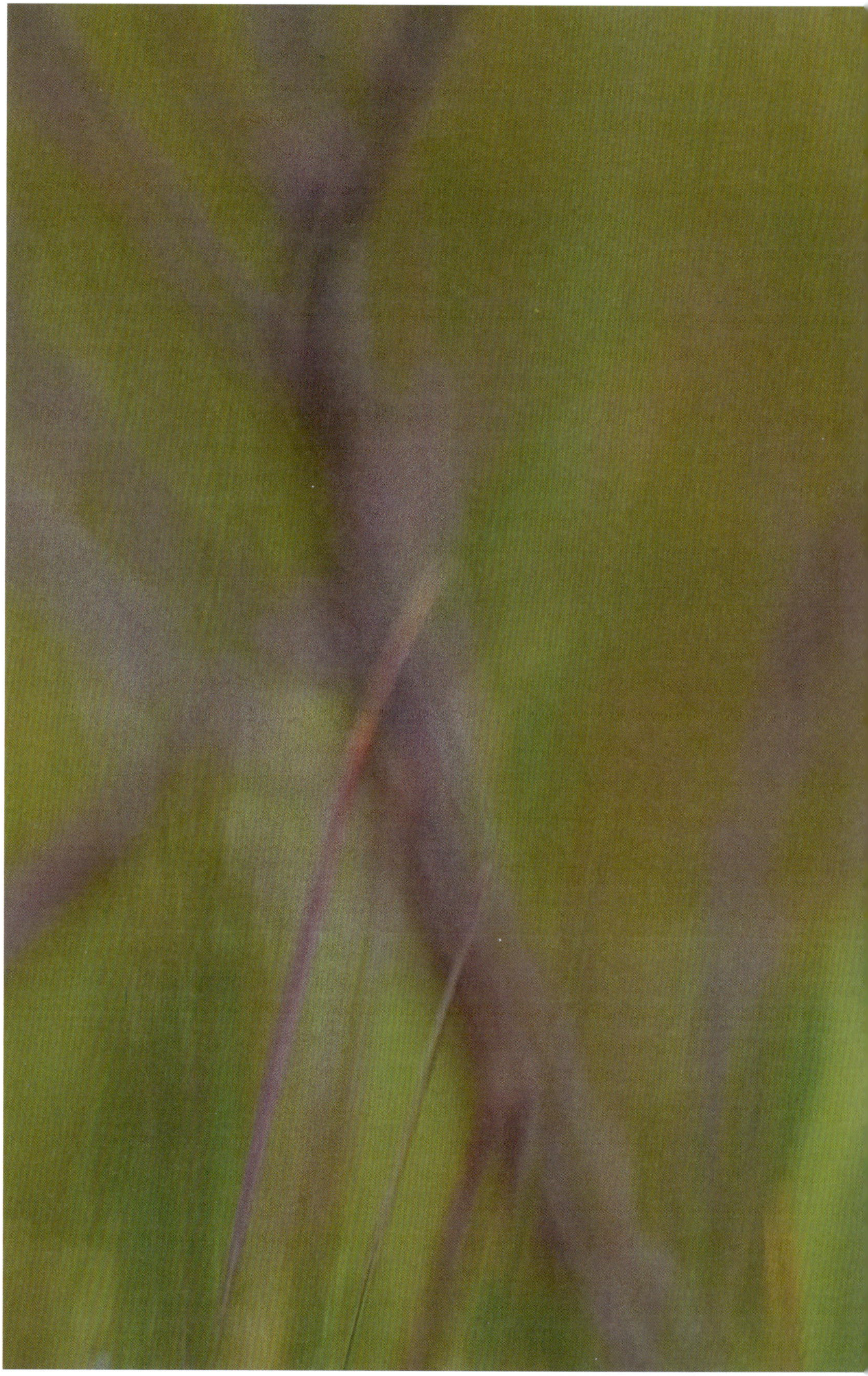

Bog asphodel

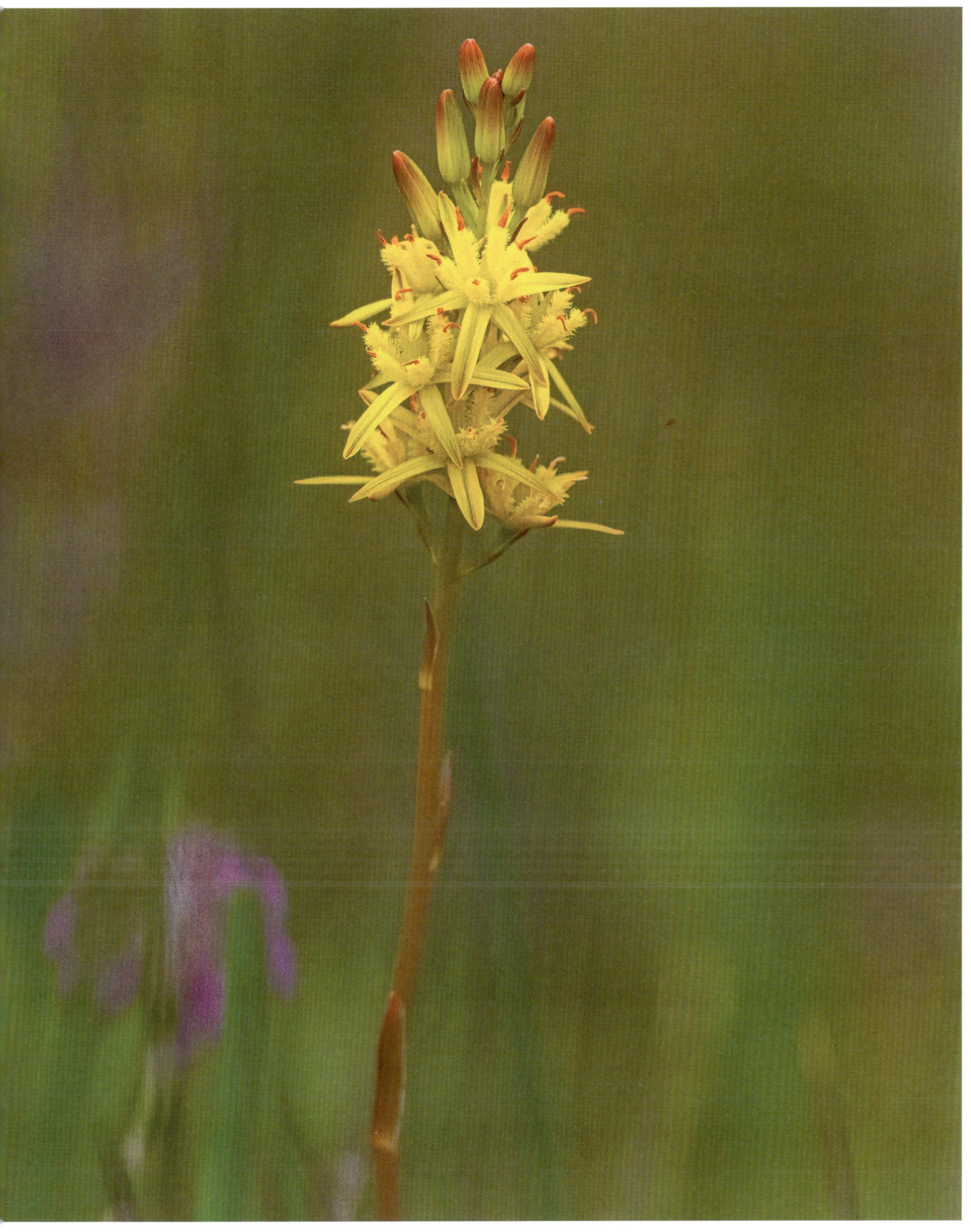

"

THE MORE CIVILISED MANKIND BECAME, THE MORE THERE WAS A PERCEPTION OF PEATLANDS AS DANGEROUS AND INHOSPITABLE PLACES

"

Ireland's eastern seaboard and the south-eastern corner in particular hold only a very small proportion of our peatlands. Only 3% of County Wexford is covered in bog, compared to over 60% of County Donegal in the northwest. The reason for this is the comparably low amount of precipitation in the east and southeast. Most of that rain falls in the mountainous areas, so that is where the majority of the region's peatlands, predominantly mountain blanket bog, can be found. The Wicklow Mountains hold by far the largest extent of blanket bog in the eastern half of Ireland, followed by the Mourne Mountains and the Antrim Plateau in Northern Ireland.

While Ireland is rich in peatlands even on a global scale, we also have one of the highest percentages of degraded peatlands. This means that the peatlands have been damaged by human activities such as peat extraction, drainage and agricultural reclamation and are no longer a living entity, no longer actively growing peat. According to the Irish Peatland Conservation Council, about 70% of Ireland's peatlands can be classified as degraded.

Once, peatlands were seen as landscapes of the mysterious and sacred, gateways to spiritual realms where one could connect with long gone ancestors and where deities and other otherworldly beings resided. Places of mystery and magic.

Bogs were also used for what is known today as paludiculture, peat-preserving agriculture on wet soil. This ranged from collecting cranberries, bilberries, the flowers of heathers, lichen and other edible flora to preserving and extending reed beds as a source for building materials. Dryer parts of the land were used for grazing livestock and making hay. The peatlands provided everything needed to make a living.

The more civilised (and I use that word hesitantly) mankind became, however, the more there was a perception of peatlands as dangerous and inhospitable places – places that were best avoided. The reasons for this unease are not difficult to understand. Even today, peatlands, especially the vast raised and blanket bogs, are hard to navigate due to their lack of distinctive features, and the soft, sticky ground makes every step an enormous effort. Peatlands are neither land nor water; firm hummocks sit beside soft, wet cushions and invisible,

Opposite top left: Evening light, Shragh Bog, County Clare
Opposite top right: Bell heather, cross-leaved heath and ling
Opposite bottom left: Overgrown turf banks and bog pine
Opposite bottom right: Cutover bog with Sitka spruce

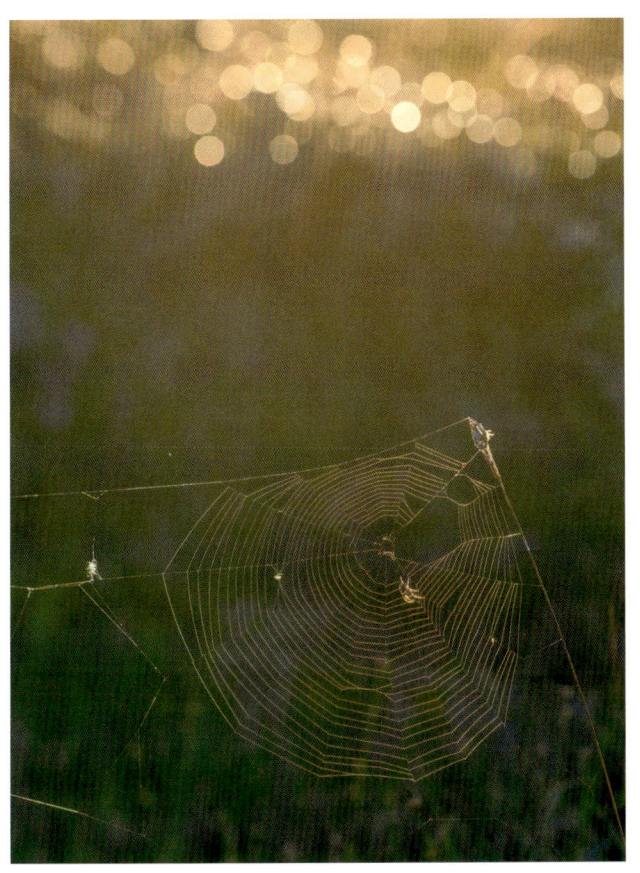

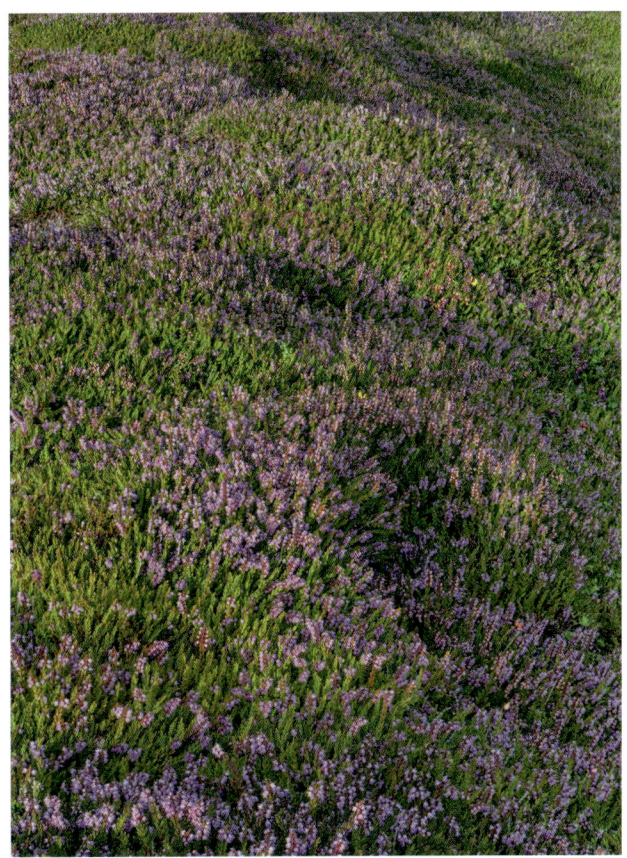

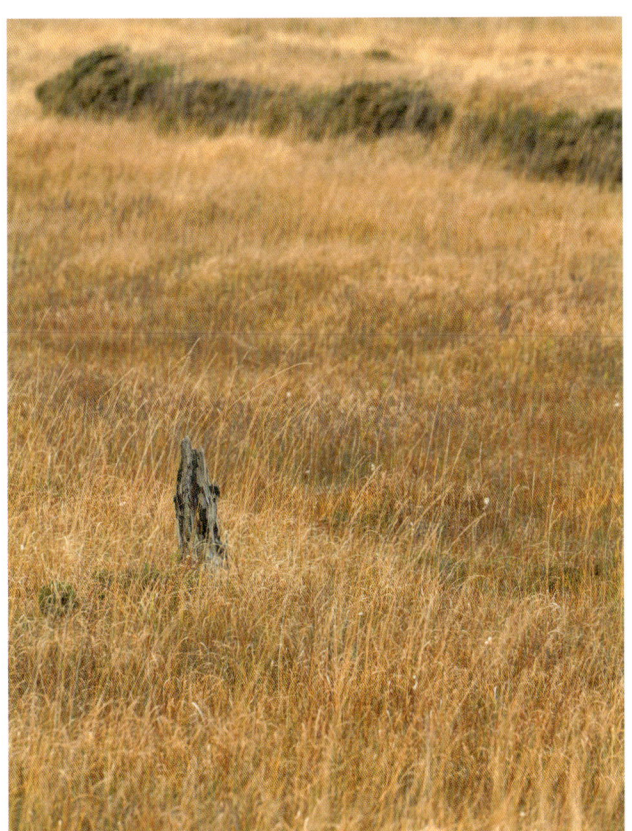

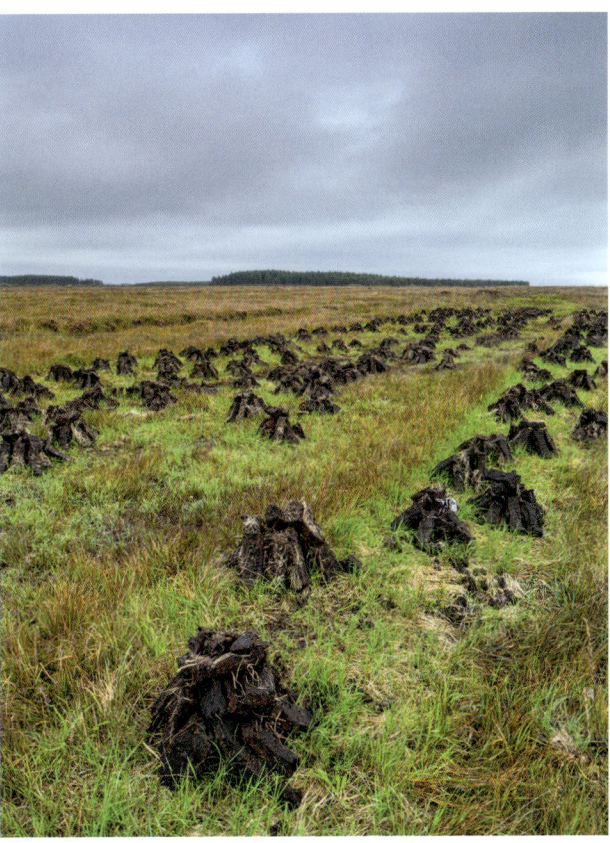

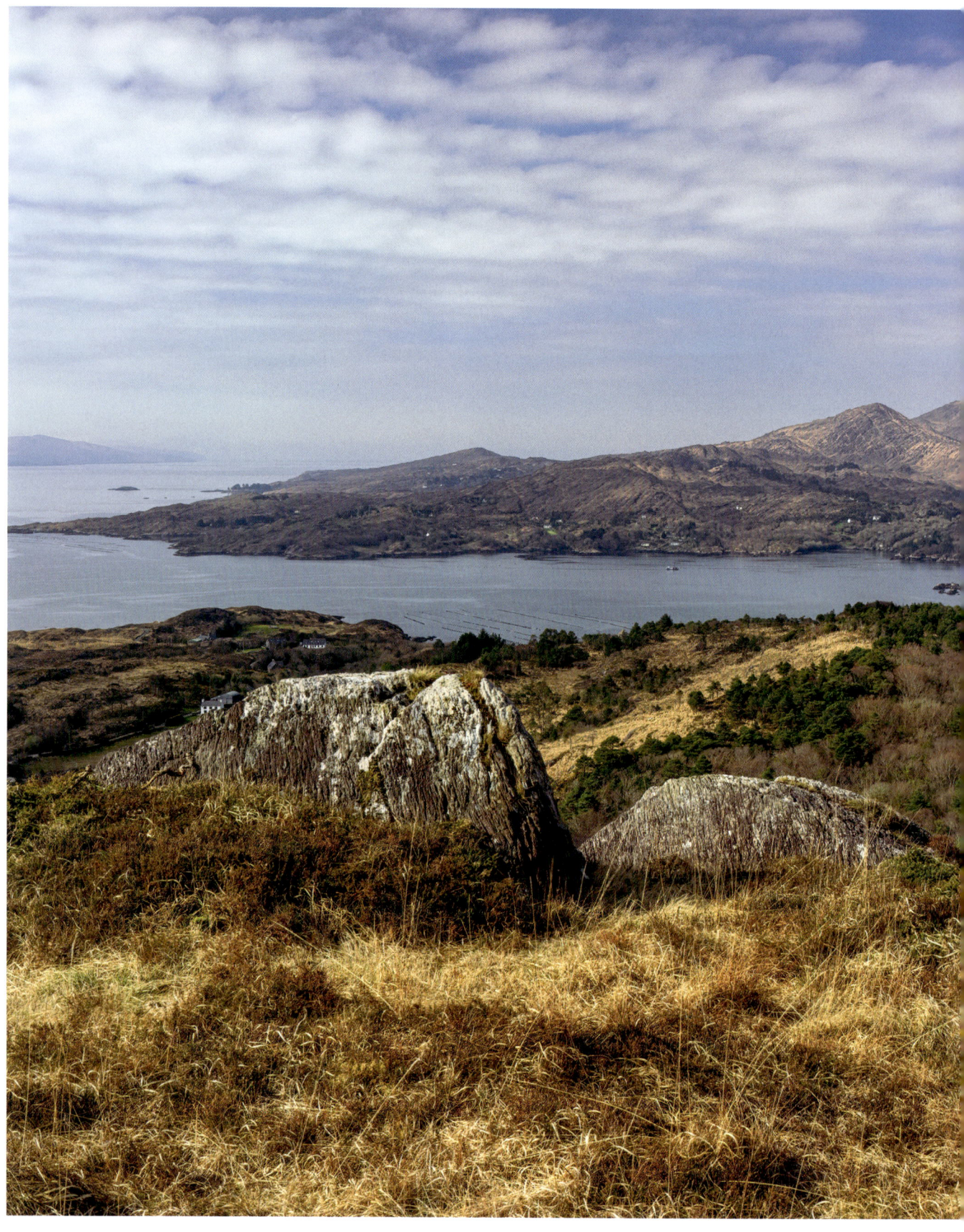

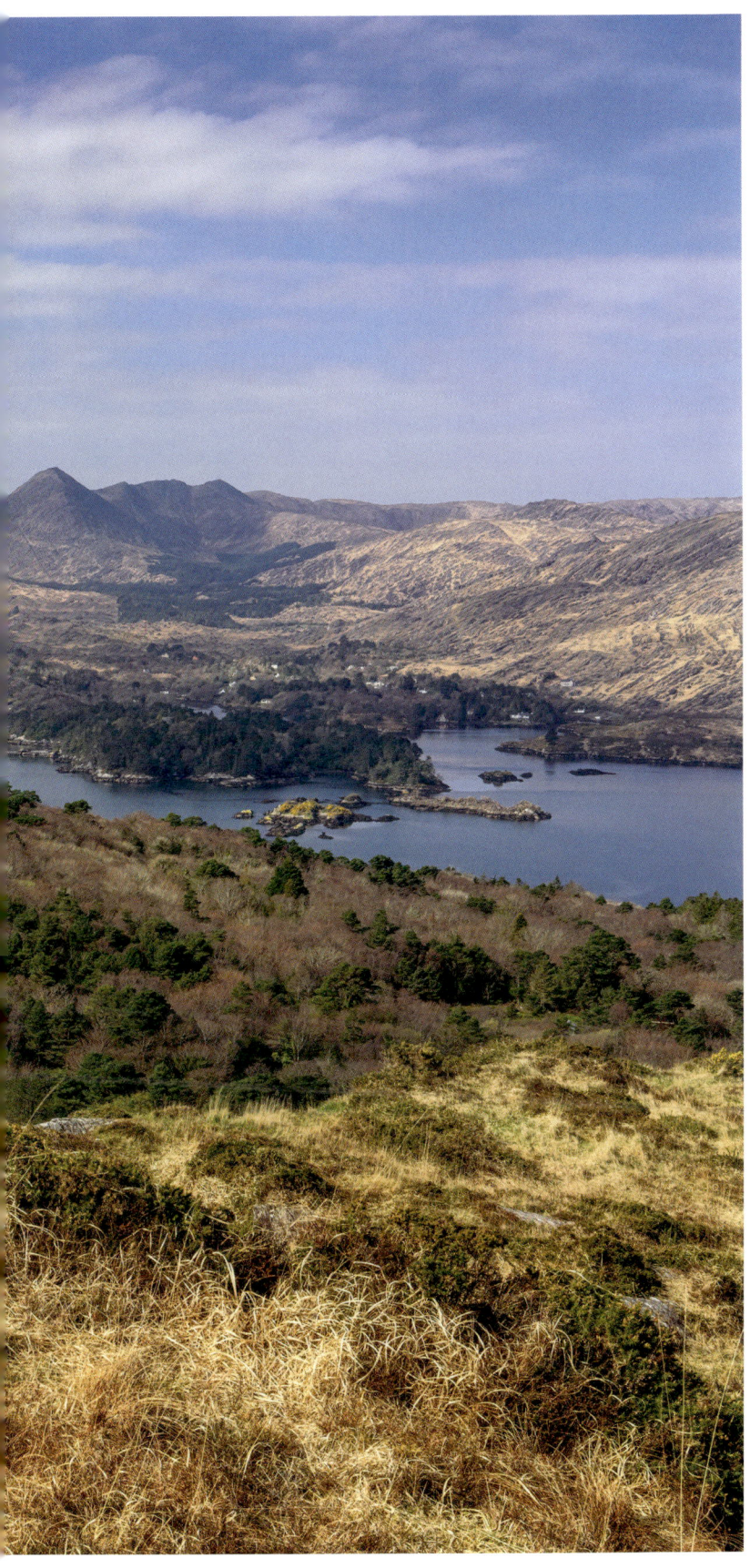

Mountain blanket bog over Glengarriff
Harbour, County Cork

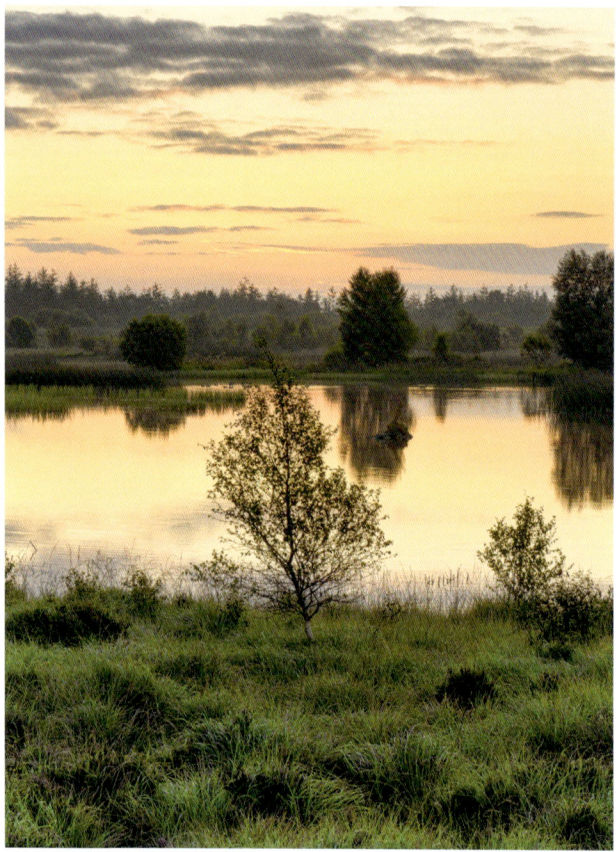

Tumduff Mor, County Offaly

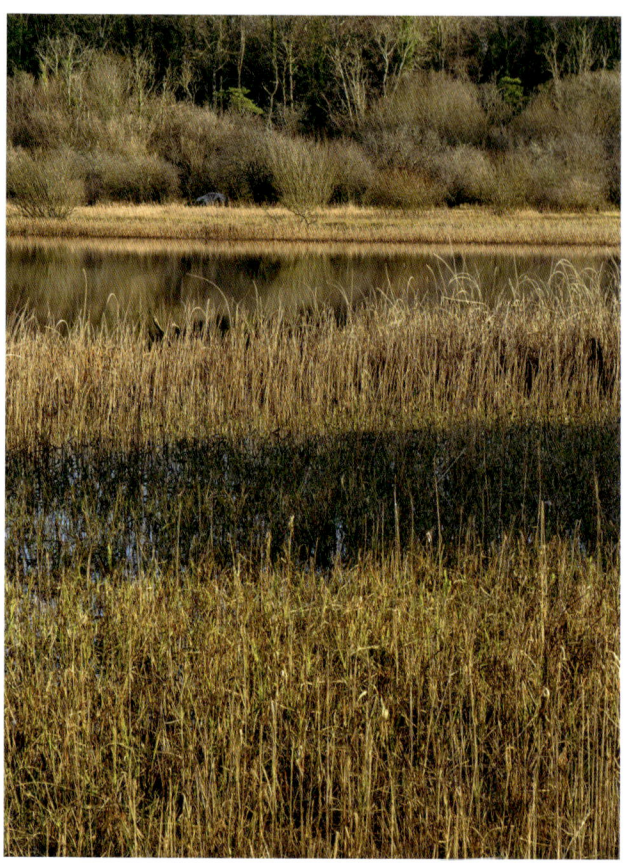

Fen landscape, Dromore Nature Reserve, County Clare

bottomless pools that could mean an unpleasant end for any inattentive traveller.

To complete the sinister picture, the slowly and incompletely decomposing surface is known to have the obnoxious odour of hydrogen sulphide, a smell closely associated with no other place than hell itself.

Another gas, the highly flammable methane, which makes up more than half of all bog emissions, has also contributed to the dark reputation of peatlands. The tale of the will-o'-the-wisps – also known as foolish fire, fairy lights, water sheeries and corpse candle – is not only known on the islands of Ireland and Britain but around the world. In the Netherlands it is known as *irrbloss*; in parts of South America (Bolivia and Argentina) as *luz mala*, and in India (West Bengal) it is called *aleya*.

All of these describe flickering lights dancing over the sodden landscape which, depending on the origins of the tale, either lead travellers safely through the bog or deliberately guide them to their death.

In Ireland, the lights are also known as Jack-O'-Lantern after Stingy Jack O'Lantern, a rather unsavoury character with a reputation as a liar, con artist, silver-tongued manipulator and drunkard. His reputation spread so far that even the devil himself heard of Jack. Equally impressed, appalled and a little envious of Jack's notoriety, Satan decided to check the infamous sinner out for himself. Should the stories turn out to be true, he planned to bring Jack with him to hell. So on one fine evening, Jack, drinking in his local pub, found himself in the presence of an inquisitive stranger. After a few words had been exchanged,

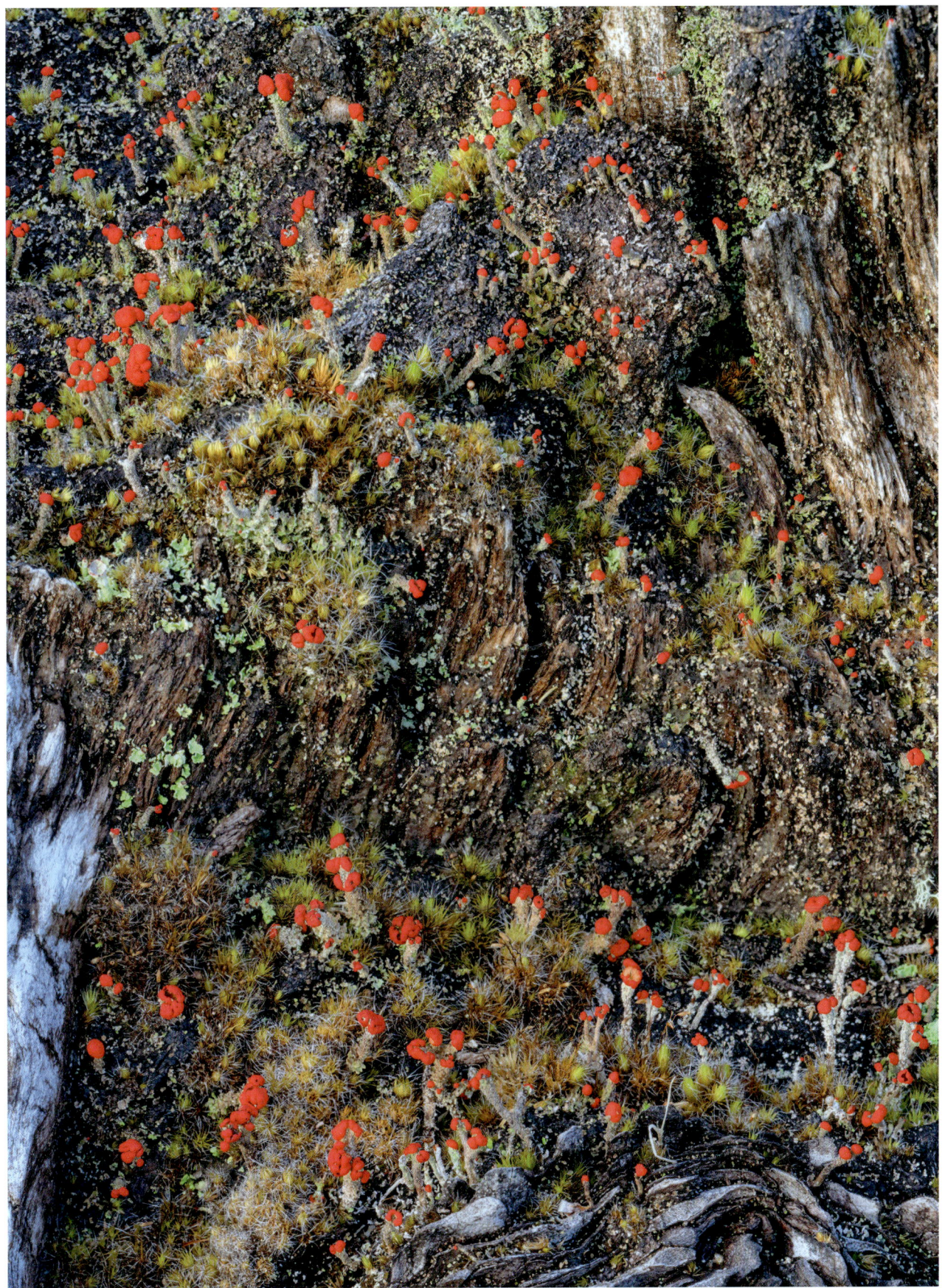

Devil's matchstick and other lichen and mosses on bog oak

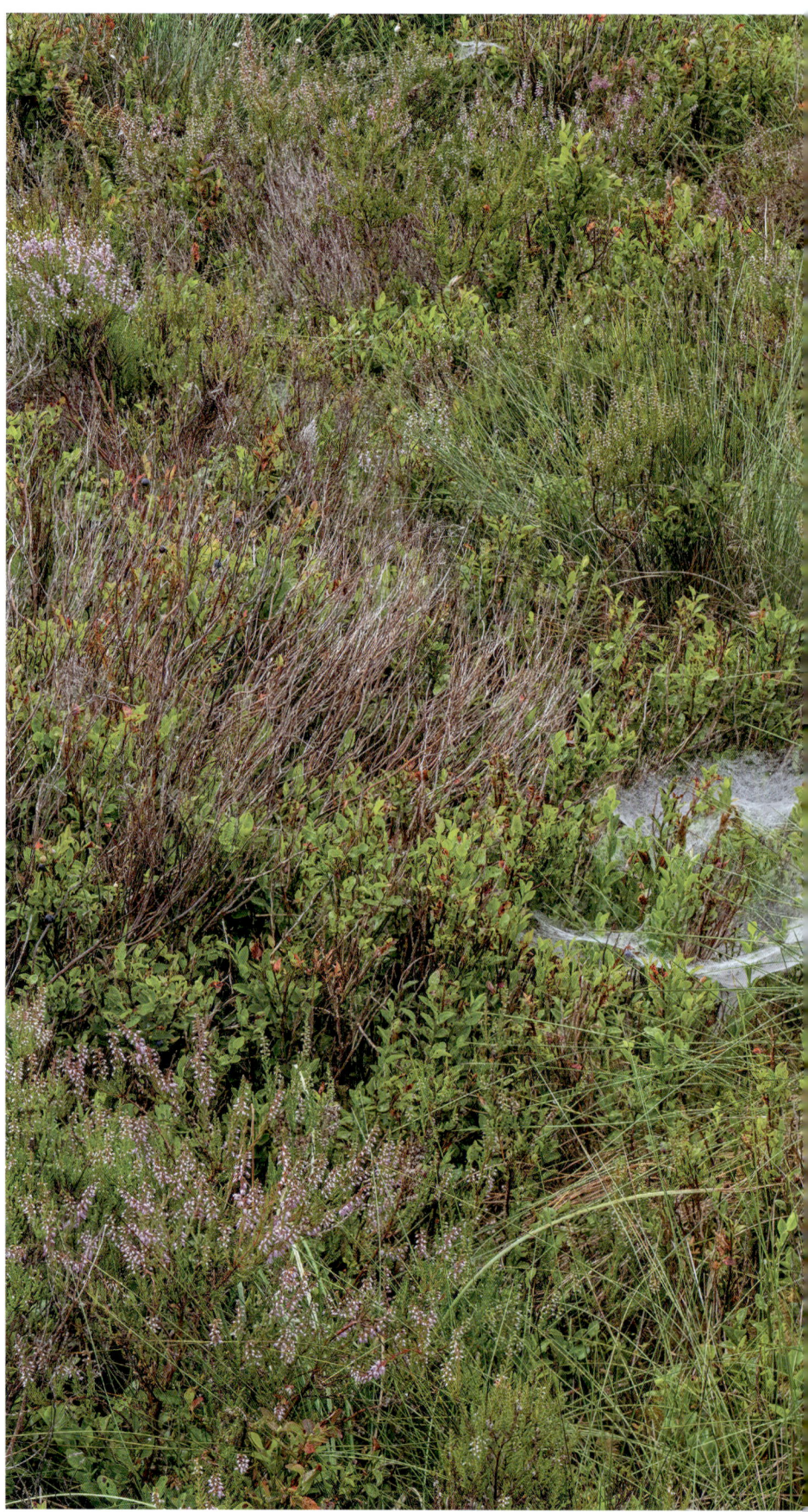

Ling and bilberry

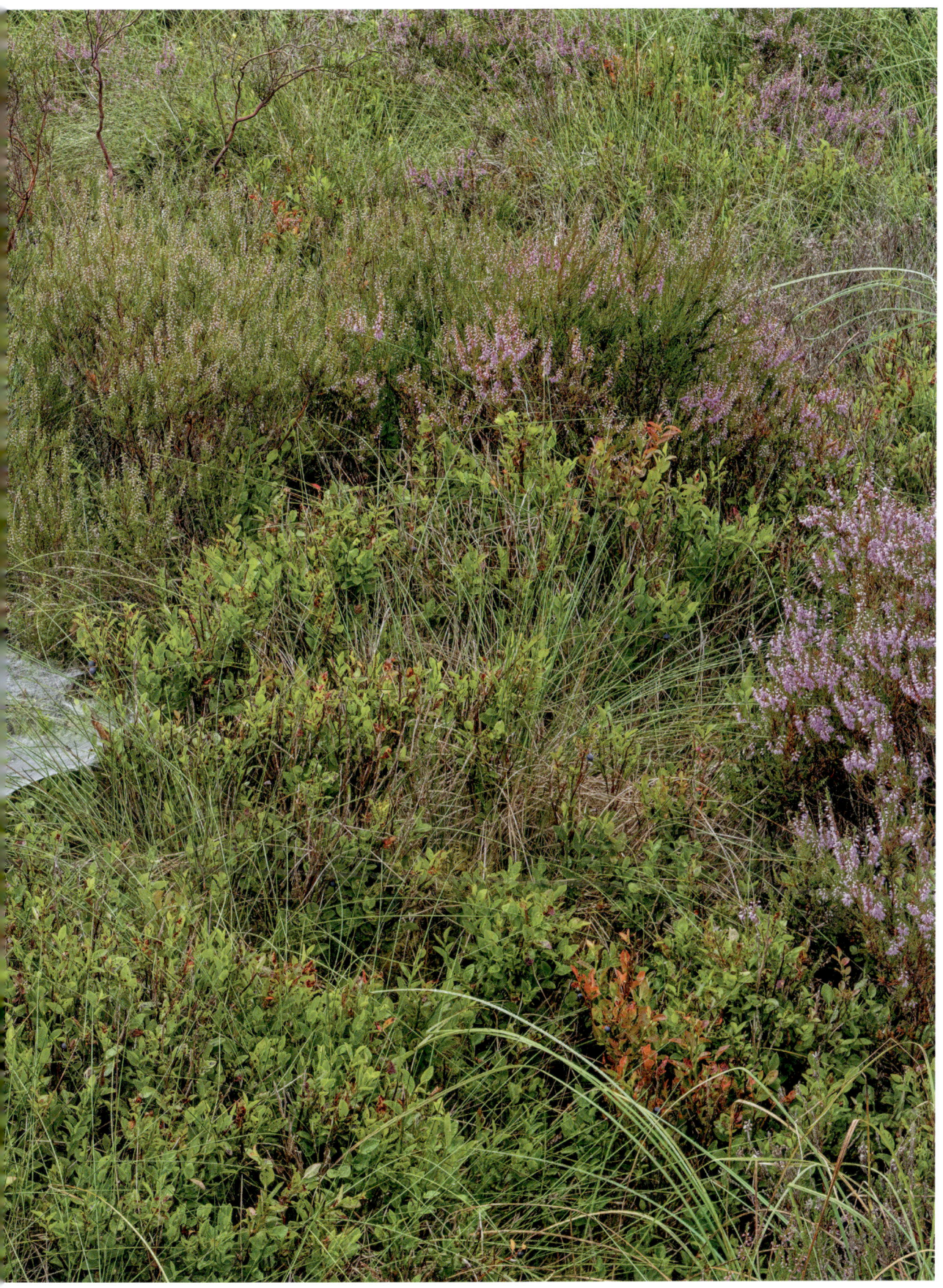

"

WILL-O'-THE-WISPS ARE NOT THE ONLY OTHERWORLDLY BEINGS FOR WHICH PEATLANDS HAVE PROVIDED AN APPROPRIATE STAGE

"

Jack realised who it was and that his life was about to come to an unfortunate end. Jack being Jack, however, did not give up easily. He persuaded the devil to allow him a few more drinks. Many hours and even more drinks later, Jack confessed that he had no money and asked politely if his new friend Lucifer would mind transforming himself into a coin so Jack could pay the bartender. The devil, a little under the influence, obliged, but instead of passing the coin to the bartender and paying his debts, Jack put the coin in his pocket. Also in the pocket was a crucifix, and Satan found himself rendered powerless. Of course this had been Jack's plan all along. Jack promised to free the raging Lord of Hell if he would grant him ten more years on earth. The devil, not having much choice, agreed, and after being freed from Jack's pocket disappeared in a rush.

The years went by and on the agreed date, the devil returned to take his price. Jack had not changed much since their first encounter and again asked Lucifer for one last wish. The devil, apparently having learned nothing, agreed. This time Jack asked for an apple from the very top of a tree. When the devil had reached the top, Jack carved a cross in the tree trunk, preventing the devil from leaving the tree. Again, Jack brokered a deal. This time he got Lucifer to accede to leave him alone forever and refrain from ever taking his soul to hell.

Many years later Jack, like all living beings, died, and his soul made its way to the gates of heaven. Not surprisingly, he was refused entry after a life of lies and deceit. Jack was sent to hell, but the devil, honouring their agreement, also refused him entrance. He did, however, provide Jack with a lantern – a hollowed-out turnip with a glowing ember from hell inside – so he wouldn't have to travel in the dark. Ever since, Jack has been wandering aimlessly, caught in the no-man's land between this world and the next.

While the dancing lights have never been fully explained, it is likely that they are either self-igniting bubbles of methane gas, fireflies or other bioluminescent insects, or simply reflections of celestial objects on the wet surface of the bog. Will-o'-the-wisps are not the only otherworldly beings for which peatlands have provided an appropriate stage. There is also the infamous shape-shifting *púca*, the screaming banshee, and a variety of fairies and bogeymen, as well as plenty

Opposite top left: Common cotton grass
Opposite top right: Bog myrtle in winter
Opposite bottom left: Bog bean
Opposite bottom right: Marsh cinquefoil

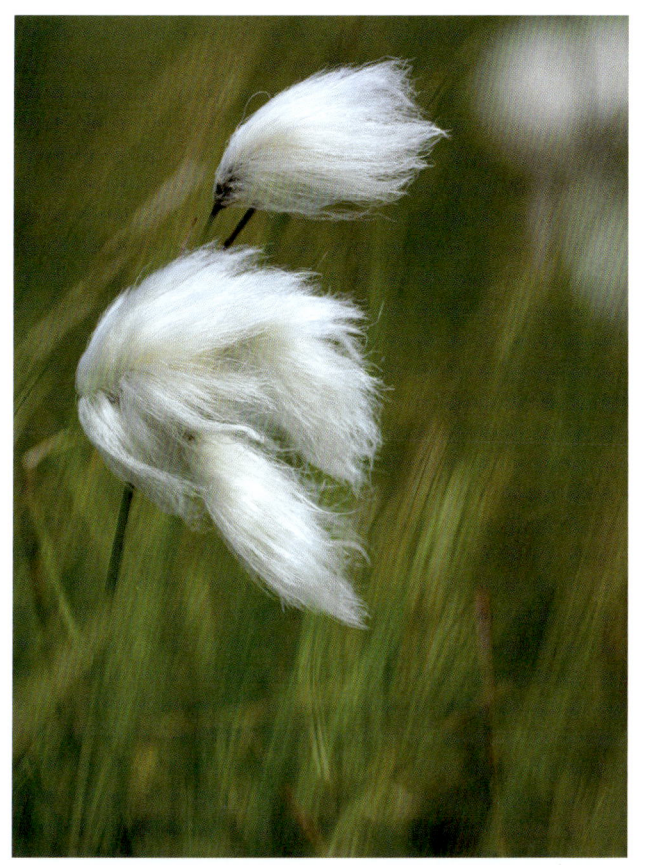

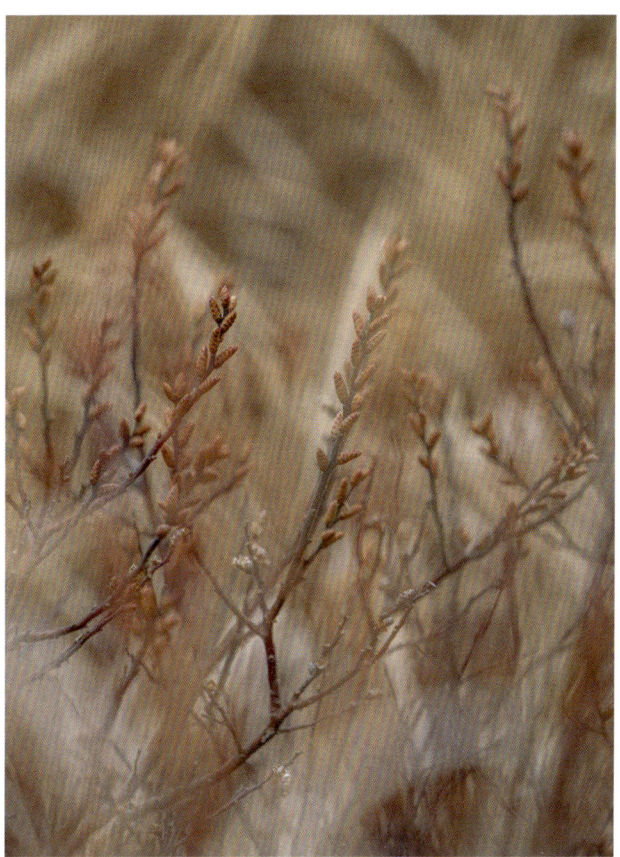

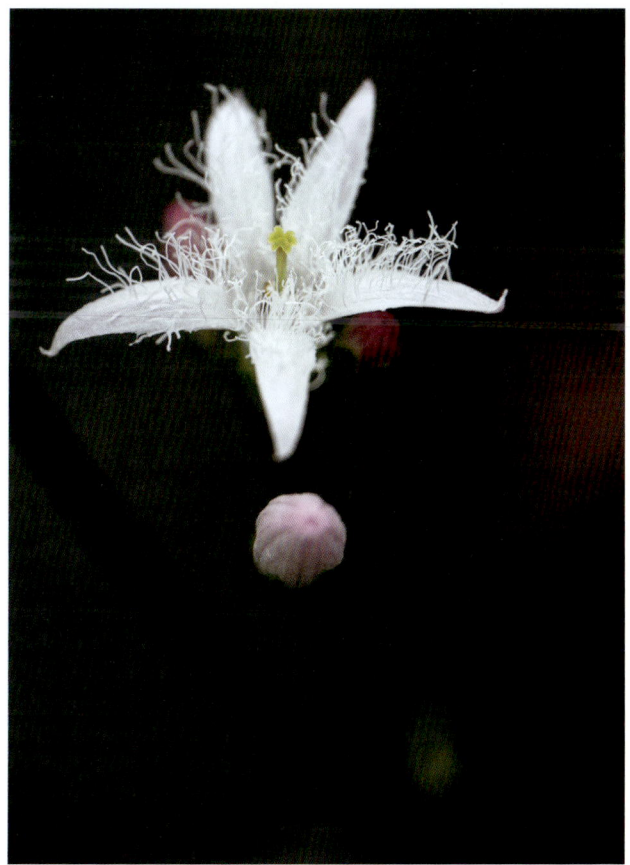

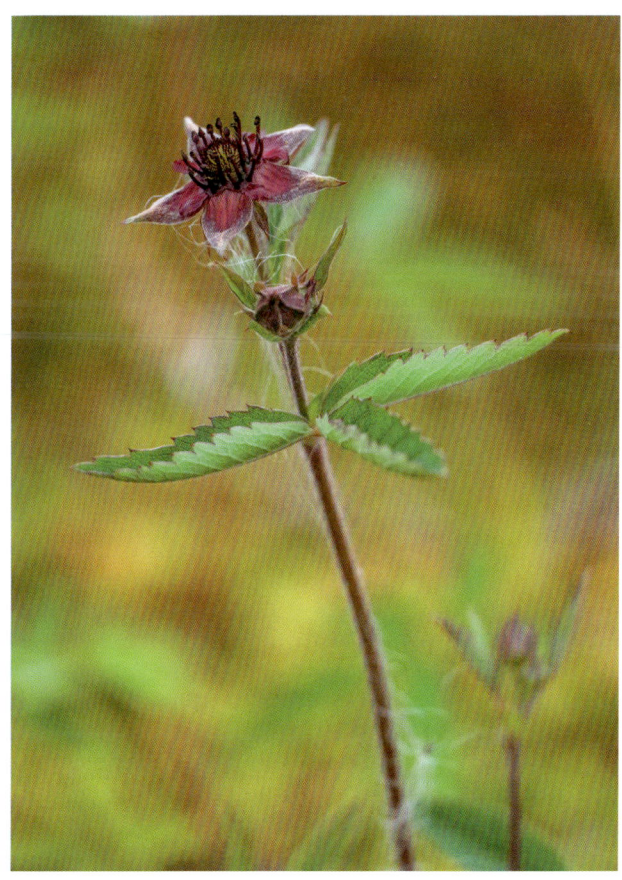

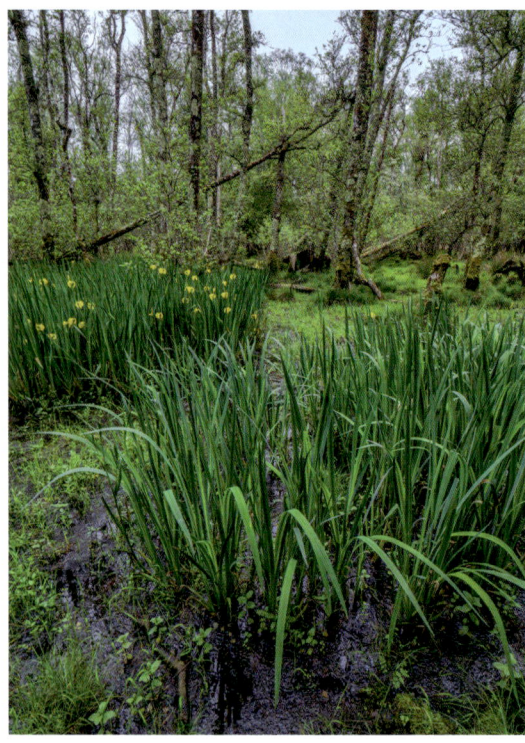

Left: Wet woodland, Ross Peninsula, County Kerry
Below: Common frog with spawn

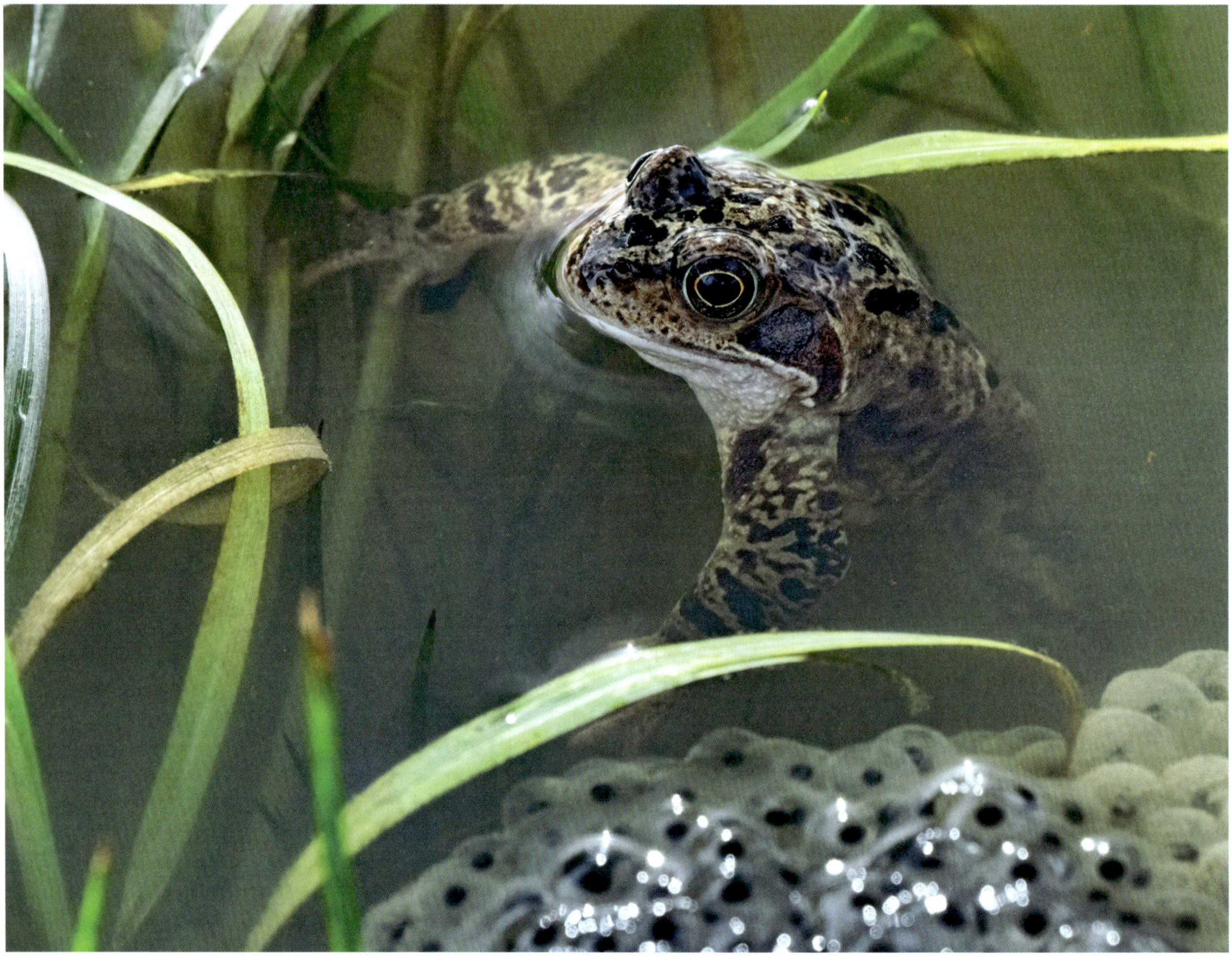

of local ghosts and spirits. The Mayo Béicheadán, or 'Mayo Yeller', for example, is a headless murder victim in the tradition of the *dullahan*, the headless horseman, that haunted his killers with mocking screams and laughter from the very bog where his body had been discarded.

The peatlands' gloomy reputation and dark perception also made their way into literature, most famously in Britain where Emily Brontë's *Wuthering Heights* and Sir Arthur Conan Doyle's *The Hound of the Baskervilles* use the landscape of the moors as a menacing backdrop. The latter claims, 'the longer one stays here, the more does the spirit of the moor sink into one's soul, its vastness, and also its grim charm'.

Back home in Ireland, the attitude towards peatlands wasn't much different. In a 12th century telling of the epic *Táin Bó Cúailnge* (*The Cattle Raid of Cooley*), they are described as 'a land of little comfort; a land that was wild, gloomy, waste, untraversable; a land covered with pools, misty, uninhabitable; a land in which foot could hardly be set, so bitter, so severe was its withering nature'. A similar sentiment was expressed in early travel writings. The Bog of Allen, which stretched over several counties in the midlands, was described as 'a vast and trackless wilderness' by Thomas Dineley in 1681. In 1698, John Dunton, an English visitor to Ireland, portrayed the Bog of Drumcullen in County Offaly as 'a treacherous and deadly place'.

Over time, the reputation of peatlands devolved from dangerous and best avoided to useless and best transformed into something productive. Peatlands were perceived as wastelands that only stood in the way of progress. With advancing engineering skills and technology, it eventually became possible to turn mires into a profitable commodity on a grand scale, a process which will be explored in more detail later in this book.

To make peatlands more useful, they were drained and turned into agricultural land or harvested to provide fuel for power stations and improve the soil of domestic gardens in the form of peat compost. As a consequence, the vast peatlands disappeared bit by bit. The few that escaped the draining and harvesting became a dumping ground for unwanted human possessions, from the broken refrigerator and the worn-out couch to the disintegrating garden furniture and rusty bicycle. Even the odd car found its last resting place in the depths of the bog. By the late 20th century, the end of thriving Irish peatlands seemed to be in sight.

Then, however, mindsets started to change. Peatlands were again revealed as places of beauty. Their function as a natural water storage facility, capable of regulating water flows and preventing flooding, was recognised. Centuries-old knowledge about the medicinal properties of peatland plants, first and foremost the sphagnum mosses, with their anti-inflammatory and antimicrobial properties, was regained. Last but not least, it was discovered that peatlands are natural carbon sinks. Today, the peatlands' biggest value lies in their capacity to remove carbon dioxide from the atmosphere and store it for eternity.

The race to restore peatlands is on, and all over Ireland projects are underway to bring bogs, fens and heaths back to life. Many of these projects are run by community organisations like the Abbeyleix Bog Project and Friends of Ardee Bog, but even semi-state institutions like Coillte and Bord na Móna have shifted their focus and now follow in the footsteps of environmental organisations like the Irish Peatland Conservation Council (IPCC) and the Irish Wildlife Trust (IWT), who have been spearheading peatland restoration projects for decades. At last, we are again starting to see what peatlands really are: places of magic and mystery.

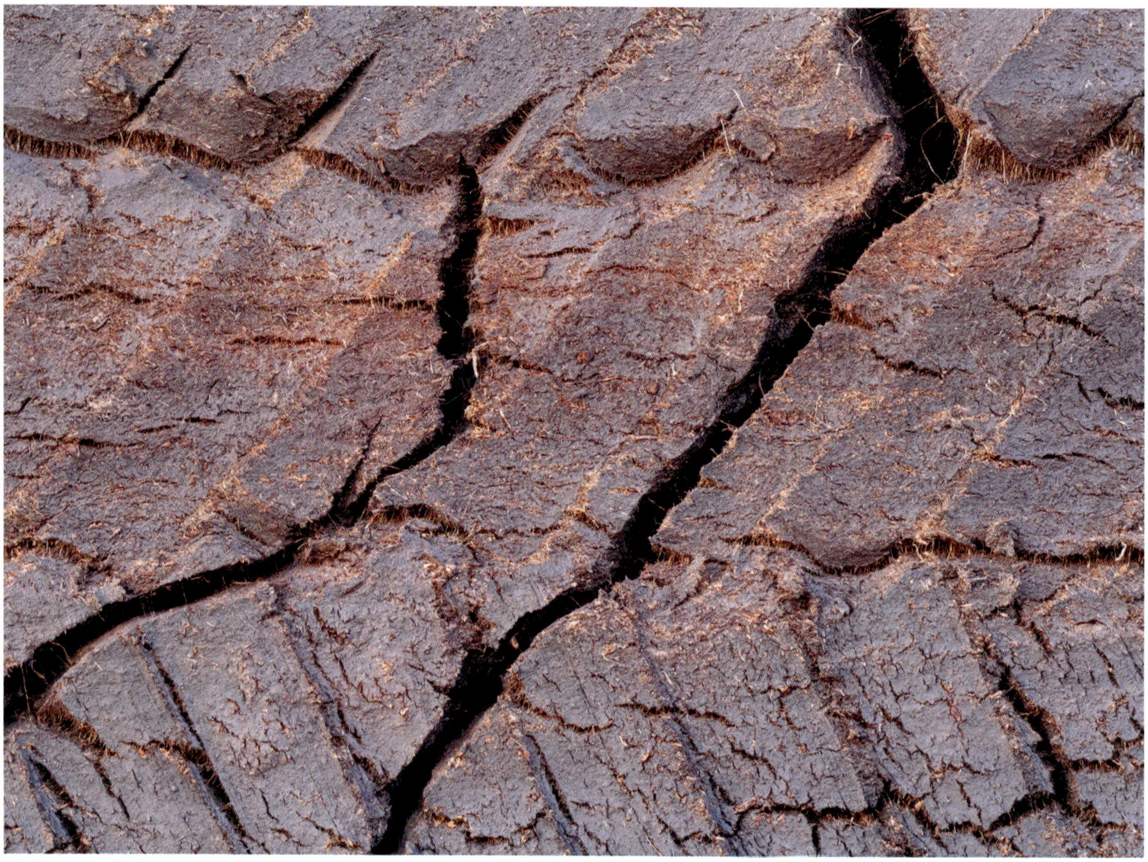

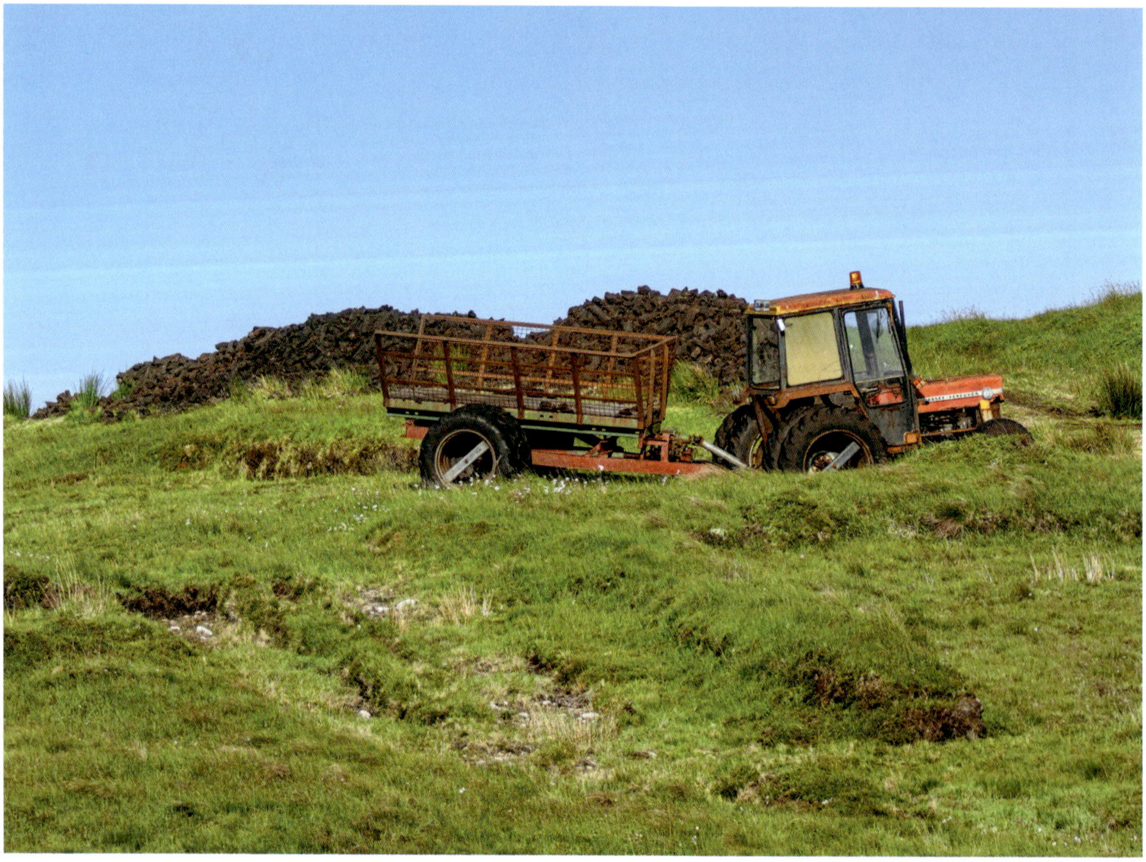

Top: Turf bank, Achill Island, County Mayo • **Bottom:** Turf stacks, Connemara

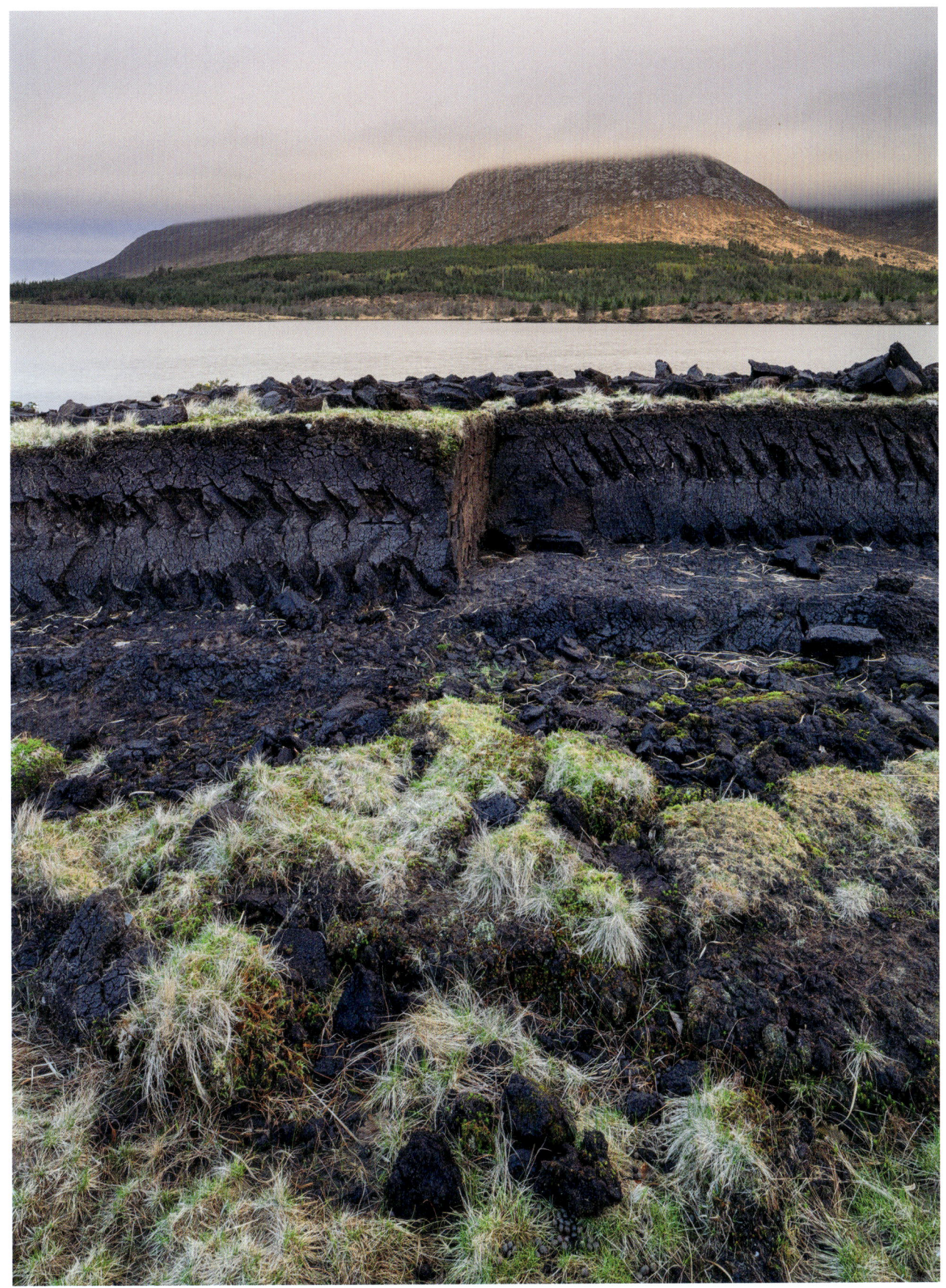

Turf bank, Inagh Valley, Connemara

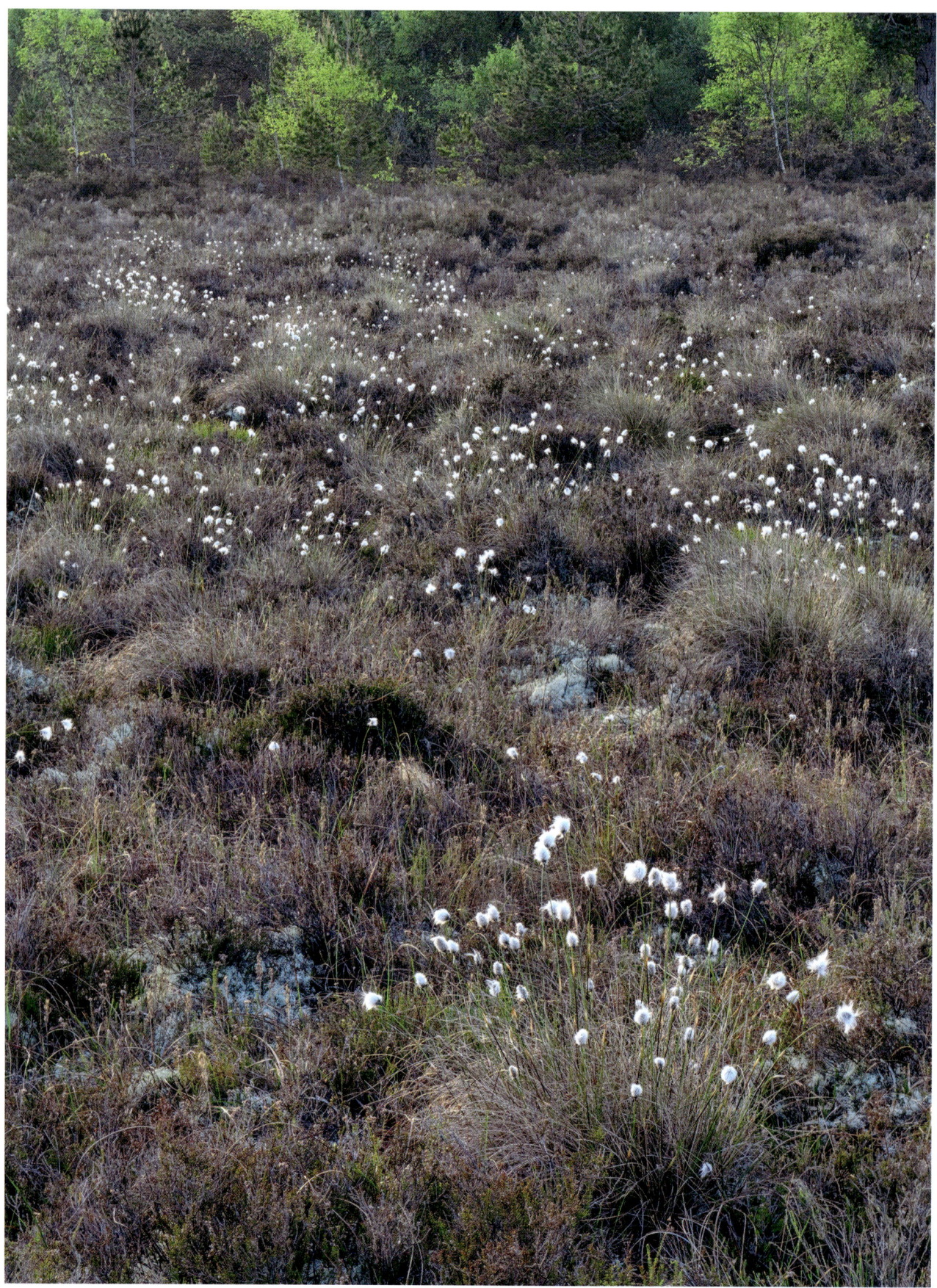

Hare's-tail cotton grass at Killaun Bog, County Offaly

Chapter Two
FROM FEN TO RAISED BOGS

Travelling through the Irish midlands today is a journey through a picture-perfect pastoral landscape. Small villages and hamlets sit within an orderly sequence of fields and pastures, separated from each other by hedgerows of blackthorn, hawthorn and the occasional mature ash, birch, rowan and oak. Here and there, small seminatural woodlands, meandering streams and tranquil lakes bring variety to what is perceived today as the quintessential Irish landscape. In the context of geological time, however, this landscape has existed for less than the blink of an eye – even a few hundred years ago, the central plain of Ireland looked very different.

The moulding of today's landscape happened during the last succession of glaciations and warm periods, collectively known as the Last Glacial Period or, more commonly, the Ice Age. The climate before the arrival of ice and snow was a warm and pleasant one. It was in this time that the blueprint for the Irish midlands was created. It is thought that back then the landscape, which was sitting on limestone bedrock covered with a layer of clay, consisted of a series of gently flowing hills and wide river valleys sprinkled with patches of steppe and forest. A network of streams, rivers and brooks provided drainage for this primeval scene.

Temperatures plummeted around 115,000 years ago, and the pristine landscape disappeared under several metres of ice, marking the beginning of the Ice Age. Over the following millennia, the glaciers advanced and retreated several times, removing the clay, grinding the underlying limestone, and in the process moving vast quantities of soil and rock back and forth.

When the ice retreated for the last time some 10,000 years ago, the landscape that was laid bare had no resemblance to its former self. Ireland's great central plain had become littered with hummocks (known as drumlins) and ridges (known as eskers) of glacial debris. The clay, sand and gravel that had been locked in the glaciers were dumped by the melting ice, completely rearranging the topography of the landscape. This rearrangement

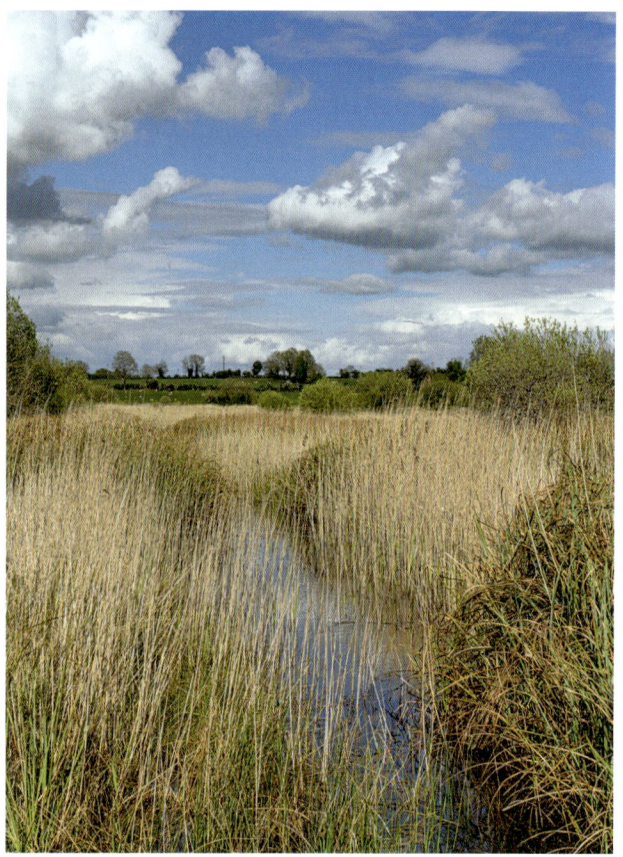

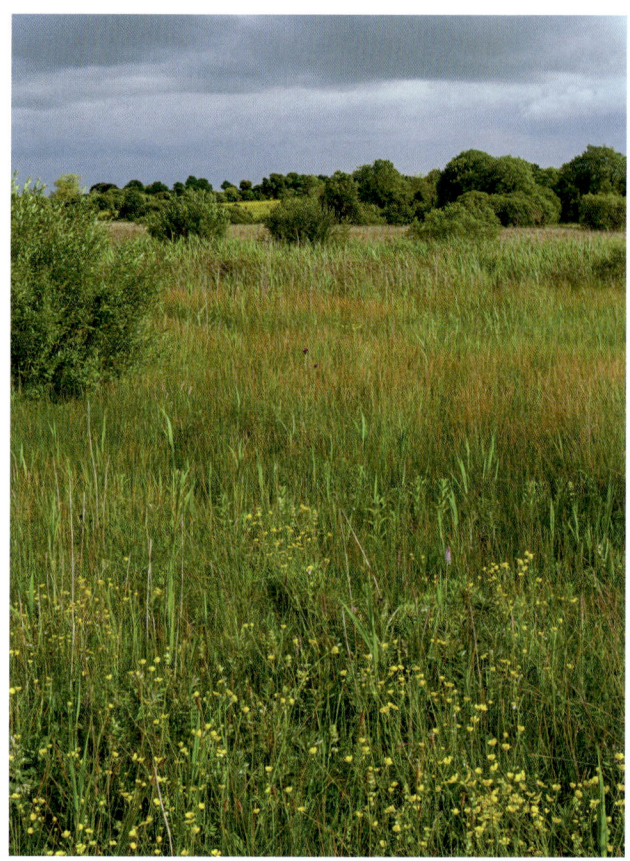

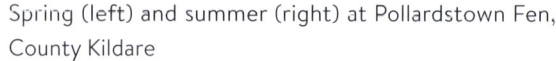

Spring (left) and summer (right) at Pollardstown Fen, County Kildare

blocked many of the old riverbeds, forcing the meltwater from the glaciers to find new ways or to stay put. Persisting frost in many places further impeded proper drainage, and so the emerging post-glacial landscape was a water world of shallow lakes, ponds and pools in between a network of eskers and drumlins.

As temperatures were rising, plants started to colonise the virgin landscape. The developing predominantly warm and dry climate was not favourable for widespread peat formation, however the shallow meltwater lakes were perfectly suited to start the process. These waterbodies supported little microbial life and were rich in nutrients – mainly calcium and magnesium, with lesser amounts of potassium and sodium – which allowed plants

like reeds, sedges and a variety of aquatic plants to flourish around and on the lakes. The first fens were developing.

Fens are peat-forming systems that differ from bogs in that they are fed by groundwater through springs and flushes or moving surface water like brooks and rivers. The nutrient-rich water these sources provide forms the foundation of an immensely diverse and luxuriant habitat. Rich or alkaline fens are the most common type of fen in Ireland. These typically develop over limestone, where the percolating water can gather large amounts of minerals and nutrients. Rather than being a uniform habitat, these fens are a jigsaw of ponds and lakes, large reedbeds, damp meadows and small woodlands, and together support a

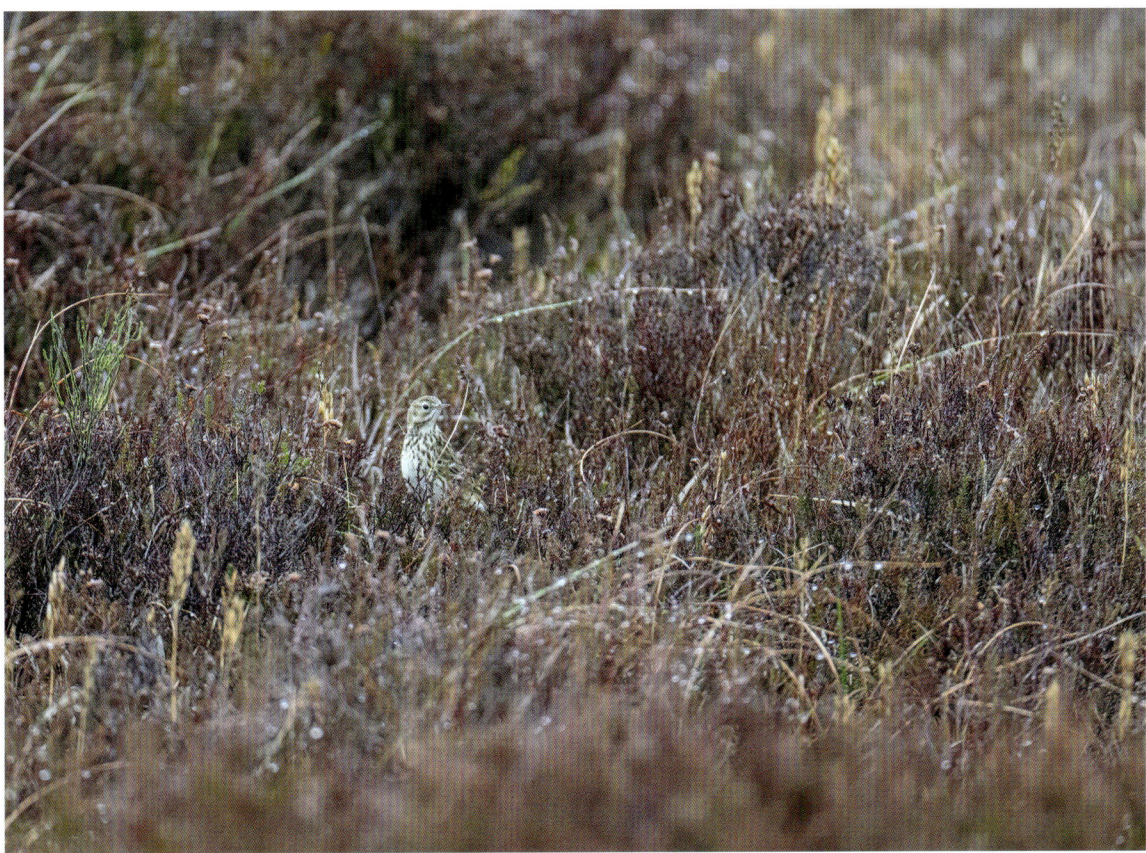

Meadow pipit at Clara Bog, County Offaly

plethora of life. The woodlands, known as fen carr, are made of alder, willow, downy birch and smaller shrubs like dogwood, guelder rose and holly. The ground under the canopy is covered in grasses, sedges and mosses, and in the wettest parts, flag iris and marsh marigold thrive. In the open outside the woodland, more wildflowers flourish, including grass of Parnassus, devil's bit scabious, marsh cinquefoil, water mint and various orchids like fly orchid and marsh helleborine. The lake margins are densely covered in bulrush and common reed, and bog bean, bog pondweed and water horsetail grow in the shallows. Unsurprisingly, this abundance attracts other wildlife. Invertebrates are numerous, including a variety of snails and insects such as dragonflies and damselflies, butterflies and moths.

This flying and crawling buffet entices birds like the meadow pipit, reed bunting and sedge warbler, who also find a cosy home in the reedbeds and woodlands. Water rail, snipe and little grebe as well as ducks and swans and amphibians like the common frog and smooth newt also feel very comfortable in the luscious environment. In total, more than 225 different plants and 635 animal species have been recorded in Irish fens.

One of the best preserved and largest calcareous spring-fed fens in Ireland today is Pollardstown Fen in County Kildare. The springs that feed Pollardstown Fen derive their water from a large underground reservoir, the Curragh Aquifer. Rainwater from the surrounding hills percolates through the limestone bedrock, picks up minerals

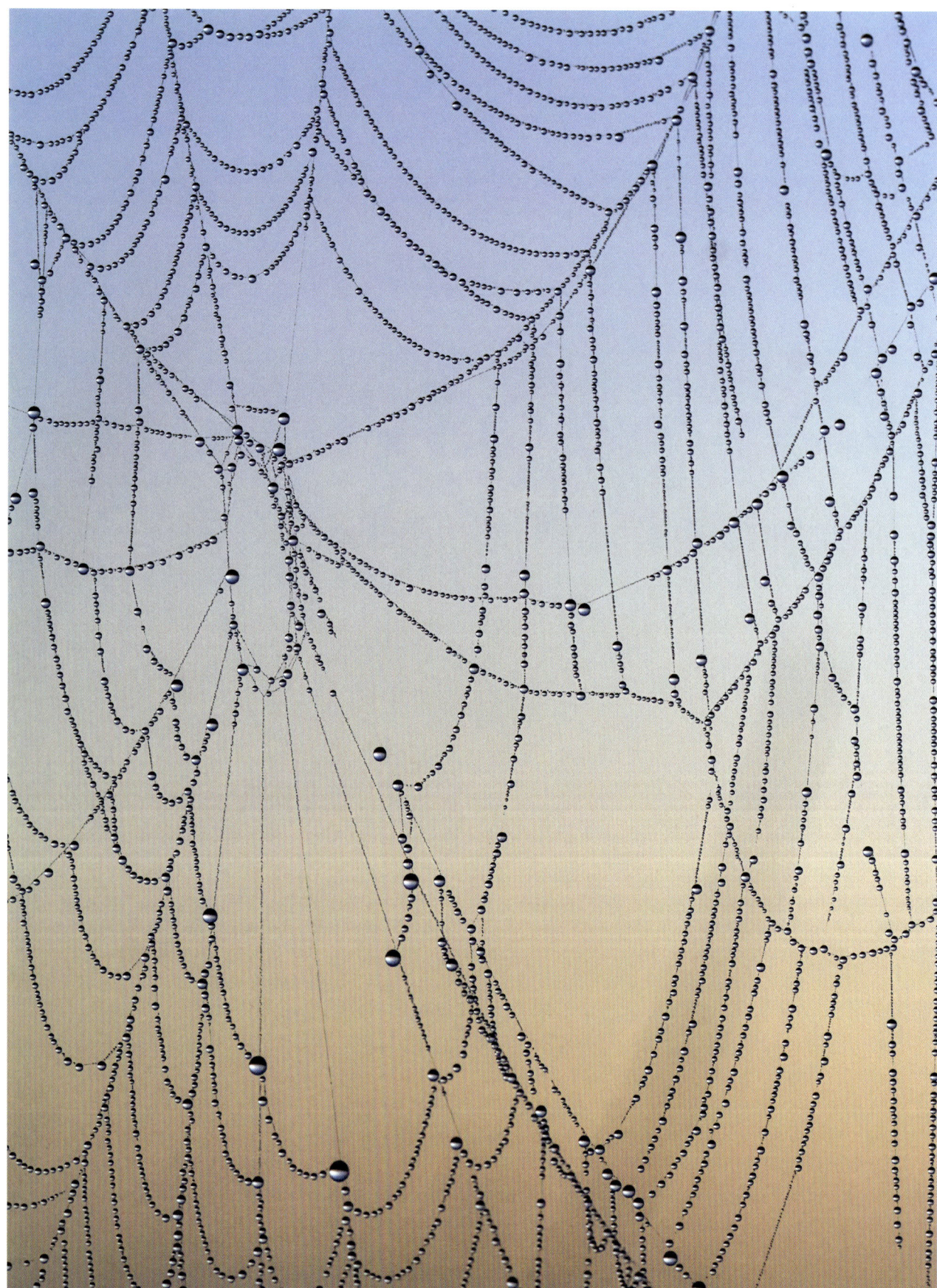

Spider's web at dawn

"

FENS LIKE POLLARDSTOWN WERE WIDESPREAD ACROSS IRELAND, BUT ONLY A FEW SURVIVED THE ANTHROPOGENIC ONSLAUGHT OF THE LAST CENTURY

"

along the way, and gathers in the aquifer, from where it moves on to eventually reach the surface, laden with nutrients, at Pollardstown Fen. Historically fens like Pollardstown would have been widespread across Ireland, especially in the midlands, but only a few survived the anthropogenic onslaught of the last century.

Pollardstown Fen covers some 220 hectares and was declared a national nature reserve in 1986. The area is a labyrinth of canals, pools and ponds, reedbeds, meadows and clusters of trees, and was made accessible to the public by a circular boardwalk. In addition to the typical fen flora and fauna, Pollardstown is known for a trio of rare whorl snails – Geyer's whorl snail, the narrow-mouthed whorl snail and Desmoulin's whorl snail – as well as some rare plant species like the narrow-leaved marsh orchid, broad-leaved cotton grass and a boreal moss called *Homalothecium nitens*, a persistent survivor from the Ice Age. Some less than common birds, the reed warbler and garganey, have found a home here, and the rare marsh harrier and savi's warbler have been spotted. Also found here are the so-called petrifying springs, which can be identified by their characteristic tufa mounds. Tufa is a porous limestone rock of a white to greyish colour that forms when calcium-rich water encounters an environment rich in carbon dioxide. This happens the moment the water from the aquifer emerges from the springs at Pollardstown. Due

to a difference in partial pressure of the carbon dioxide in the water and the carbon dioxide in the air, some of the carbon dioxide in the water is released. This causes a rise of pH in the water, which causes calcium carbonate to precipitate and settle around the spring, over time forming the unmistakable tufa mounds.

Poor or acid fen is rare in Ireland. In appearance, poor fens are similar to rich fens, but they support a considerably smaller variety of plants – mainly grasses, sedges and mosses – and also host specialists that are more commonly associated with bogs, like the carnivorous common butterwort and sphagnum mosses. The reason for this is that poor fens are fed by nutrient-poor water sources, which can be found in areas with underlying acidic rocks like granite or rhyolite. Poor fens can also be a transition stage between rich fen and raised bog after the rich fen loses its water source, which can happen for a variety of reasons: streams and rivers can change their course; wells can dry up; or extensive peat growth can form a barrier between the surface and the water source.

The formation of peat in the fens started as early as the immediate post-glacial period. Remains of dead plants sank to the bottom of the lakes, and because of the lack of microbial life, these remains only partly decomposed; they accumulated over time, and subsequently the first layers of fen peat were formed.

As this process progressed, the edges of the water bodies filled in, permitting plant communities to encroach. As reedbeds extended bit by bit and the peat layers piled up, the lakes grew smaller and shallower. Eventually the peat reached the surface, filling in the lakes completely. Another contributing factor was the rebound of the land after the weight of the ice had been lifted. While the land rose up, water levels in the lakes dropped, accelerating the filling in of the lake with peat. At this stage, not only did the lakes cease to exist, the peat layers would have reached a thickness that prevented the plants on the surface from reaching the groundwater table below. The rich fens had become poor fens, and as a consequence the entire plant community changed. Plants that relied on nutrient-rich groundwater disappeared and were replaced by species that were able to thrive under much starker conditions; mainly sedges and brown mosses like *Scorpidium scorpioides*, *Drepanocladus revolvens* and *Calliergon giganteum*. These mosses would then have laid the groundwork for another fundamental change, the change from an alkaline to an acidic environment. Once the brown mosses had done their job, the masters of the bog, the sphagnum mosses, could colonise the juvenile raised bog, marking the starting point for a complete change in living conditions and the transition from fen to bog.

Sphagnum mosses are known as the bog builder, which is particularly true for raised bogs. Worldwide, around 380 different sphagnum species are known, and in Ireland twenty-four species have been identified. The visual differences are minute, but each species is dedicated and perfectly adapted to a particular area of the bog. There are the hummock builders, the red bog moss (*Sphagnum capillifolium*), Austin's bog moss (*Sphagnum austinii*) and rusty bog moss (*Sphagnum fuscum*); the carpet formers are papillose bog moss (*Sphagnum papillosum*) and magellanic bog moss (*Sphagnum medium*); and some species even spend their entire existence submerged in bog pools like the feathery bog moss (*Sphagnum cuspidatum*) which, due to its appearance when removed from the water, is also known as drowned kitten moss.

Sphagnum mosses are not only intriguingly beautiful and mesmerisingly well adapted to their habitat, but these plants also go a big step further: they are ecosystem engineers and create their own living conditions, adapting the environment to their needs and in the process making life challenging for other species.

Before this happens, however, the sphagnum mosses first must get to the soon-to-be raised bog. How plants colonise a new space is always a bit of an enigma and relies to some extent on chance. Sphagnum mosses, like all mosses, proliferate

Opposite top: Lough Boora, County Offaly
Opposite bottom left: Frogbit
Opposite bottom right: Sphagnum and grass

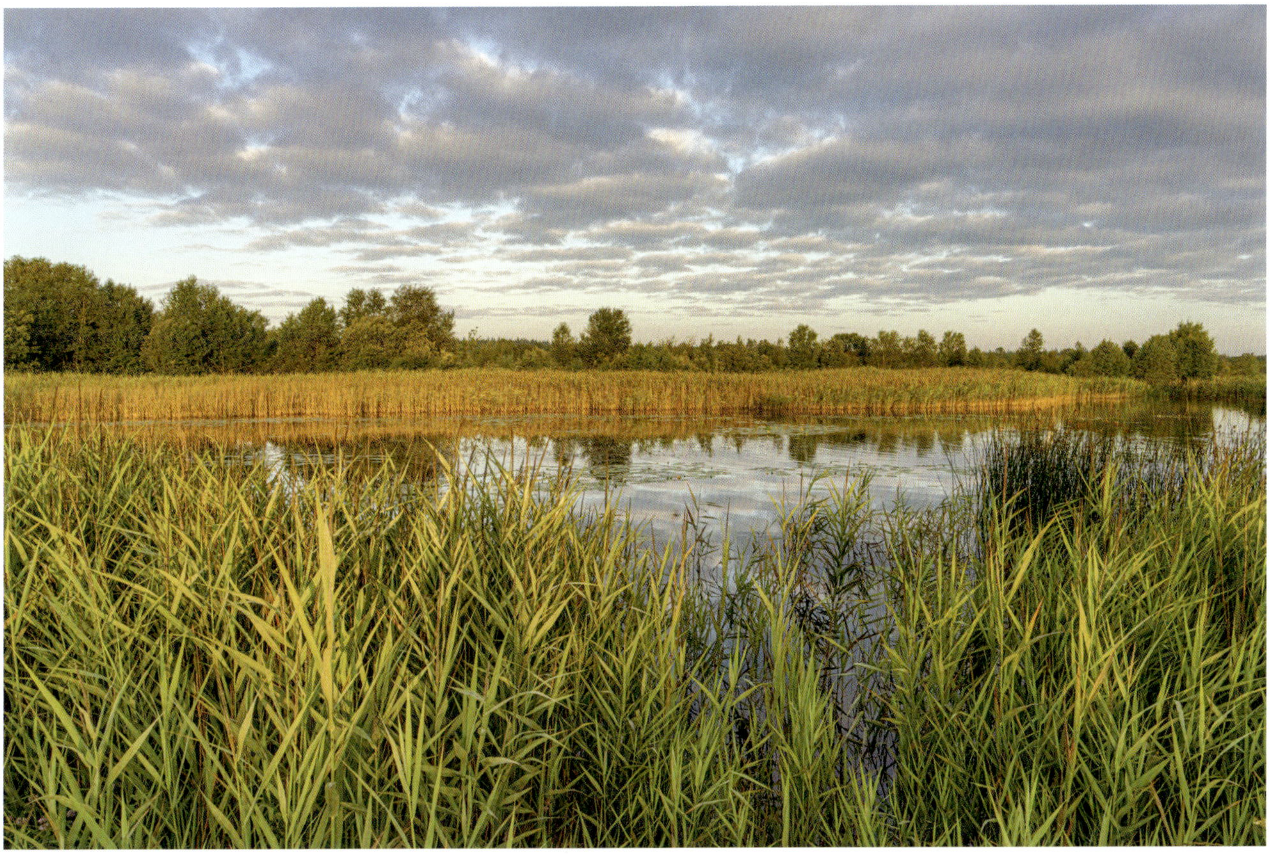

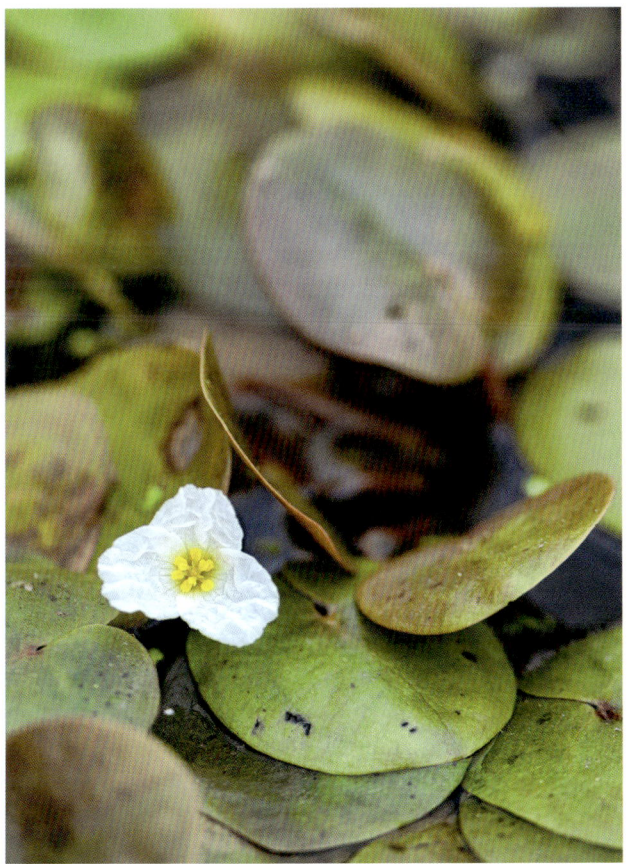

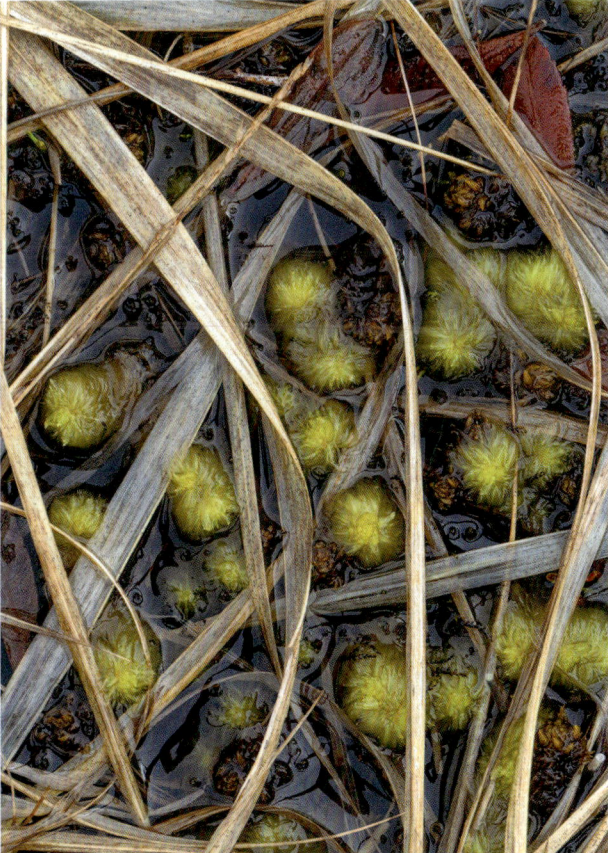

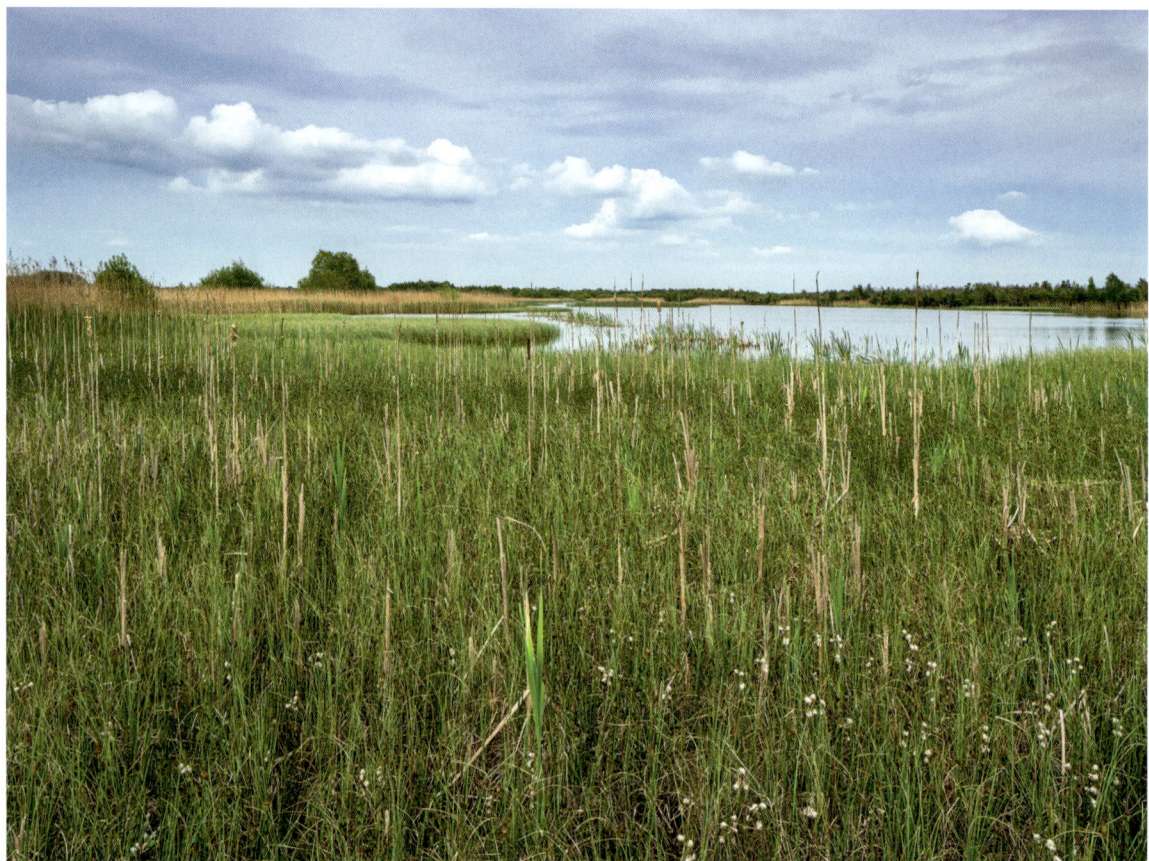

Summer (above) and winter (opposite) at Turraun Wetlands, County Offaly

through spores. These are tiny entities, carried away from the parent plant by wind. Should one or more of these spores land on a suitable surface, they start to grow and form a new colony.

Once they have established themselves, the sphagnum mosses start to make the place more comfortable by altering the chemical balance of their new home. Because mosses have no roots, they need to extract nutrients from their immediate environment. Producing uronic acid in their cell walls allows the plants to release hydrogen ions into their surroundings, which triggers a process that attracts nourishing calcium and magnesium cations into the mosses. This exchange, however, reduces the pH around the plants, creating the acidic environment the sphagnum mosses love – but few other plants can cope with.

The sphagnum mosses don't stop there. In addition to uronic acid, sphagnum mosses also generate other acidic compounds called phenolics with names like sphagnan, sphagnol and sphagnorubin. These compounds have different responsibilities while inside the plant: some strengthen cell walls, others act as a sun block and give the red colour to some sphagnum species. When these compounds leach into the environment, however, they kill off microbial life and as a side effect also give the water of the bog ponds and pools its characteristic molasses-brown tint. The antibacterial properties of the phenolics are an important component in peat formation. The new acidic conditions, in combination with the plummeting number of microbes, on top of the already anoxic conditions caused by waterlogging, slow the decomposition rate con-

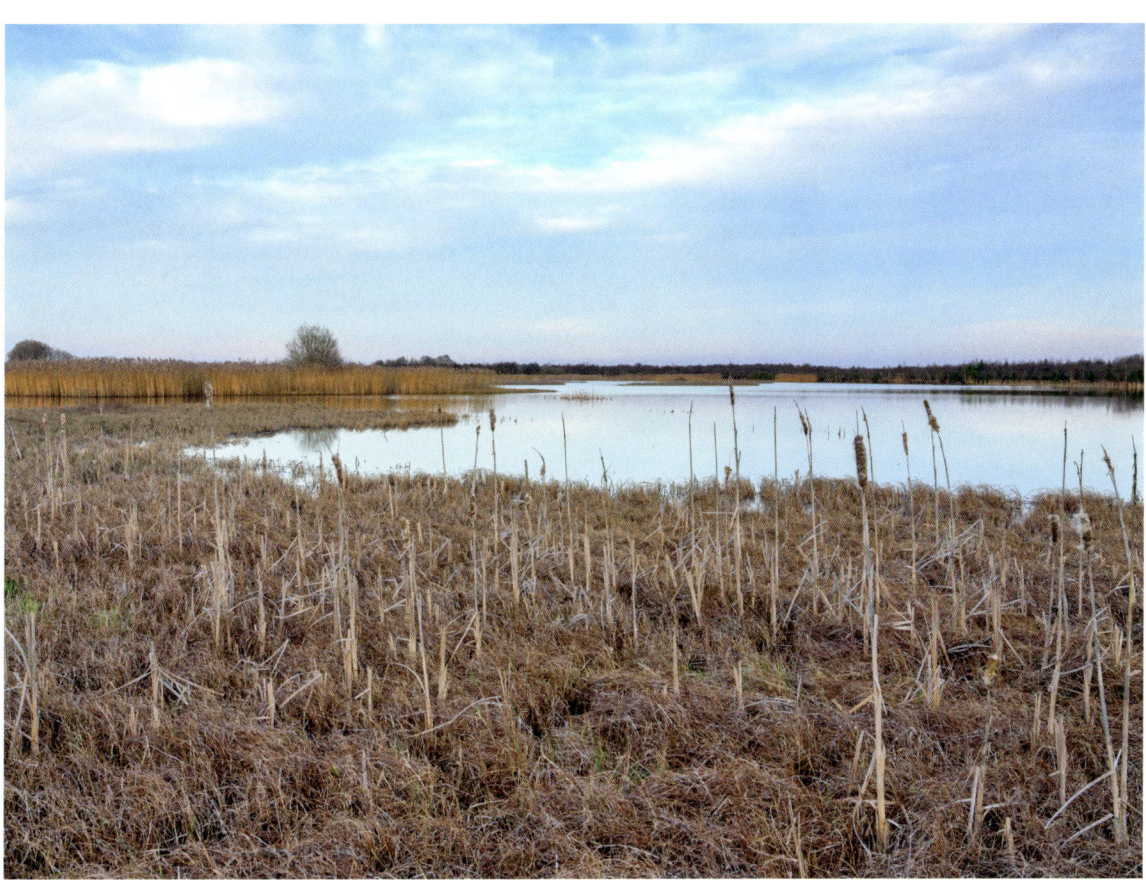

siderably and trigger further accumulation of peat. These newly formed peat layers consist mainly of the remains of sphagnum plants and are of a much lighter colour and softer texture than the rather dark and fibrous peat of the fen.

Altering their environment through chemical processes is not all sphagnum mosses can do. They also have a rather peculiar morphology that gives them further advantages. All sphagnum mosses consist of a main vertical stem on top of which sits the apical meristem, the part of the plant responsible for upward growth. The stem itself produces branches that are short and tightly packed around the upper part of the stem, longer and less dense further down. These branches are attached to the stem in clusters known as fascicles, and the number of branches in each fascicle is a distinguishing feature of the different sphagnum species. Each fascicle contains two types of branch, one hanging down and the other spreading vertically.

Both branches and stem are covered in small leaves that are extremely thin, consisting of only one cell layer made up of two cell types. The chlorophyllous cells perform photosynthesis and are more numerous and considerably bigger than the hyaline cells, which are responsible for water storage. It is the latter, in tandem with the arrangement of the leaves, that give sphagnum mosses the ability to store multiple times their own weight in water, thereby making an active contribution to the waterlogged condition of the bog. In addition, sphagnum mosses can transport water through capillaries from the lower parts of the plant to the top, an ability that keeps raised bogs wet even

"

THE BOG DOMES IN IRELAND
ARE FLATTER AND LESS DEFINED
THAN ON THE EUROPEAN
MAINLAND

"

during prolonged dry periods, at least for a while. The density in which the individual mosses grow – depending on the species, one to seven individual plants can occupy a square centimetre – creates thick carpets, which further helps in retaining moisture.

Keeping the top of the plant moist is crucial to the sphagnum moss's ability to photosynthesise. To transform light energy into chemical energy, sphagnum mosses need to be well drenched – photosynthesis works best when the upper part of the plant is at a water saturation of 700%. Light levels fall extremely quickly under the sphagnum canopy, and the lower parts of the plants don't get enough sunlight to perform photosynthesis effectively. These parts, which extend for several centimetres, are, however, still alive and can regrow the plants should the top be damaged.

From a depth of about ten centimetres, the stems and their branches are dead, embedded in newly formed peat but still connected to the living upper part of the plants. This connectiveness is the sphagnum mosses' secret weapon. Should nutrients become scarce on the surface, the mosses can translocate reserves stored in the dead part of the plants to the upper part where they are needed.

In the first millennia after the glaciers had disappeared, the climate settled into a warm – sometimes hot – and dry period, comparable to that of today's Mediterranean. These conditions were not very favourable for peat development, and the subsequent formation of fens and bogs was slow and limited to only a few suitable places. This changed around 4,000 years ago when Ireland's climate transitioned to the Atlantic maritime regime of the present. The reasons for this change are not entirely clear but were part of a more widespread climate shift in the northern hemisphere that marked the end of a global warm period known as the Holocene climate optimum.

The new climate brought cooler summers, milder winters, and a considerable increase in precipitation. These new conditions were exactly what Ireland's juvenile peatlands needed to thrive and expand. Peat growth exploded; fens transformed into raised bogs, which grew, layer upon layer, into their characteristic dome shape, dictated by the upward growth pattern of the sphagnum mosses. Ireland's raised bogs are, however, different from their counterparts on the continent. The bog domes in Ireland are flatter and less defined than on the European mainland. Our raised bogs also have a distinct flora; most notable in Ireland's oceanic version of the raised bog is the lack of trees.

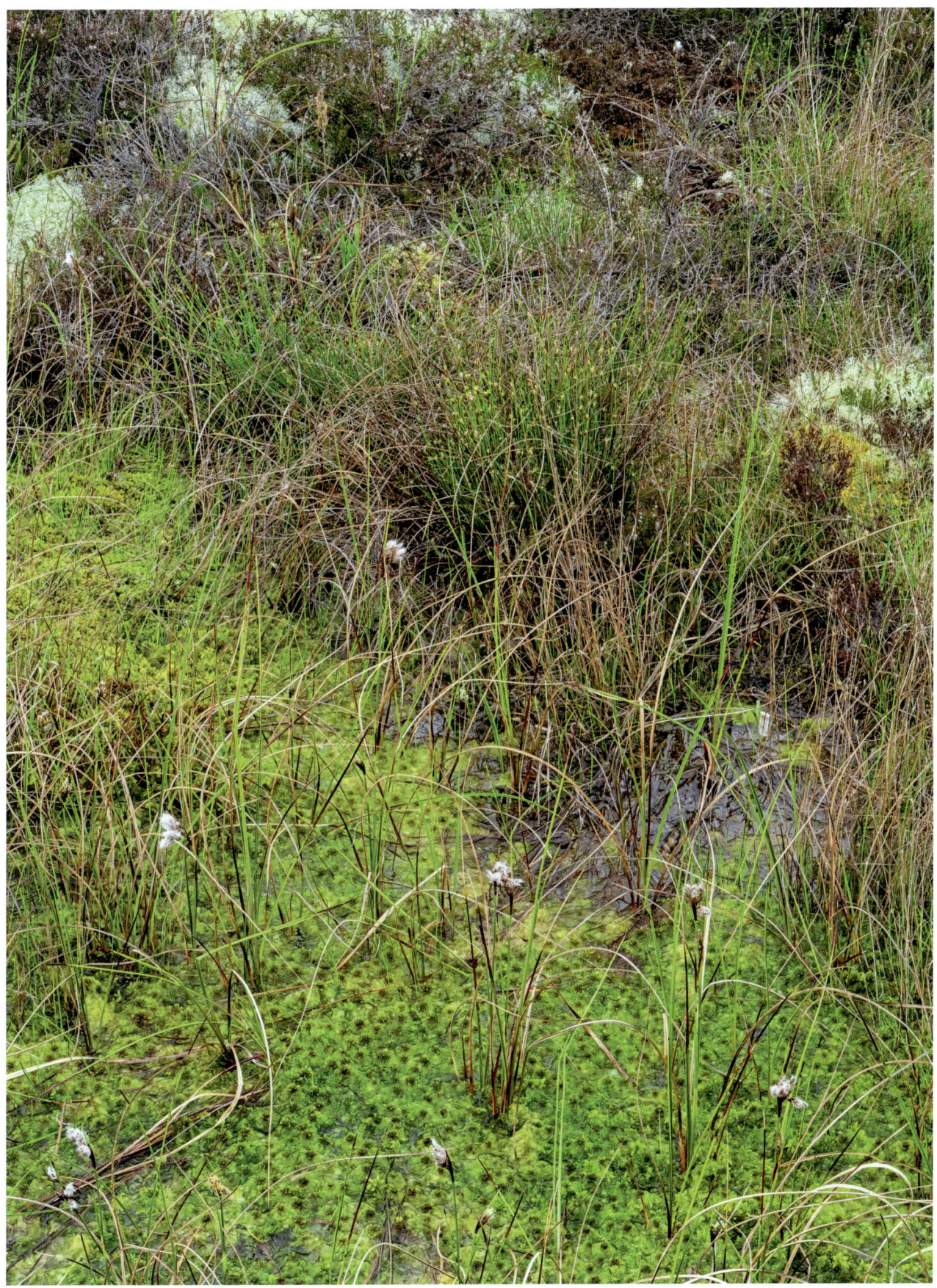

Bog pool, Ferbane Bog, County Offaly

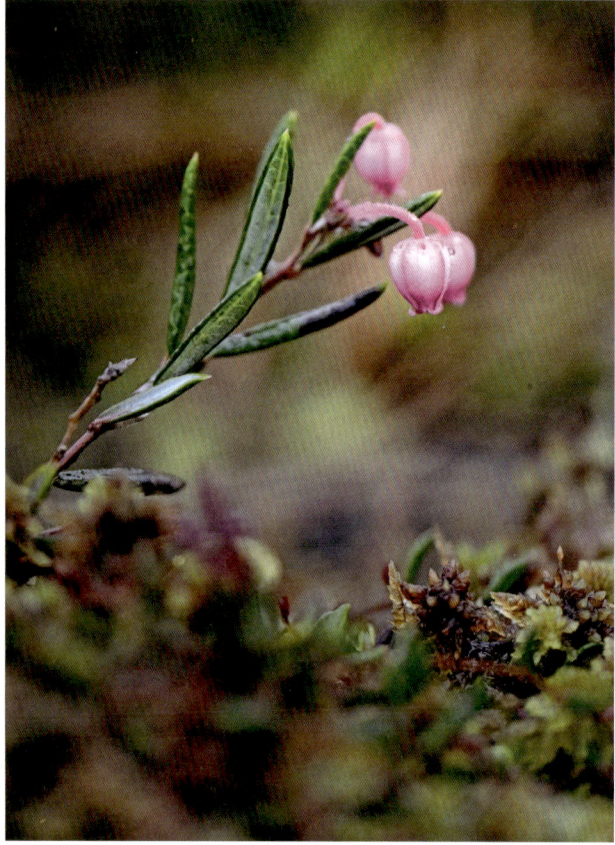

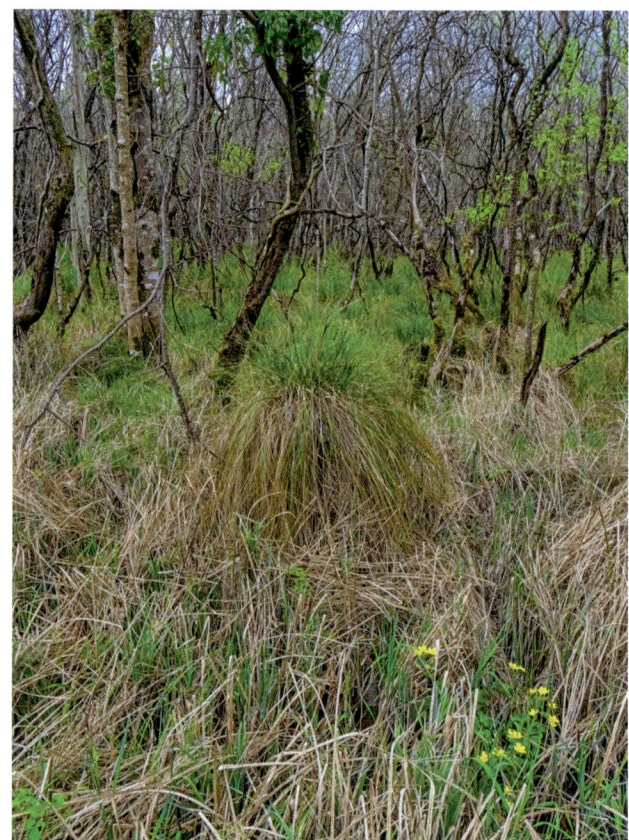

Above left: Bog rosemary
Above right: Wet woodland with marsh marigold at Abbeyleix Bog, County Laois
Opposite top left: Bog asphodel and cross-leaved heath
Opposite top right: Harvested raised bog, County Offaly
Opposite bottom left: Round-leaved sundew
Opposite bottom right: Bog bean

Even within Ireland, different versions of the raised bog can be found. Looking at raised bogs west of the midlands and comparing them with their relatives in the centre of the country, a change in flora becomes obvious. The western raised bogs are missing some of the most familiar flowers of the midlands' raised bogs, cranberry and bog rosemary, and show greater numbers of grasses and sedges. Sometimes these bogs are referred to as intermediate bog, because they seem to contain elements of both raised and blanket bog. The reason for this split personality can be found in the precipitation of the different locations. Typical raised bogs can be found in areas with between 750 and 1,000 millimetres of precipitation per year, while intermediate bogs thrive in places where rainfall accumulates to 1,000 to 1,250 millimetres per year, which is more likely in the western half of the country.

As the individual raised bogs grew and expanded, they eventually joined with neighbouring peat domes, and so the large raised bogs of the midlands came into being. Today, with only a few small plots

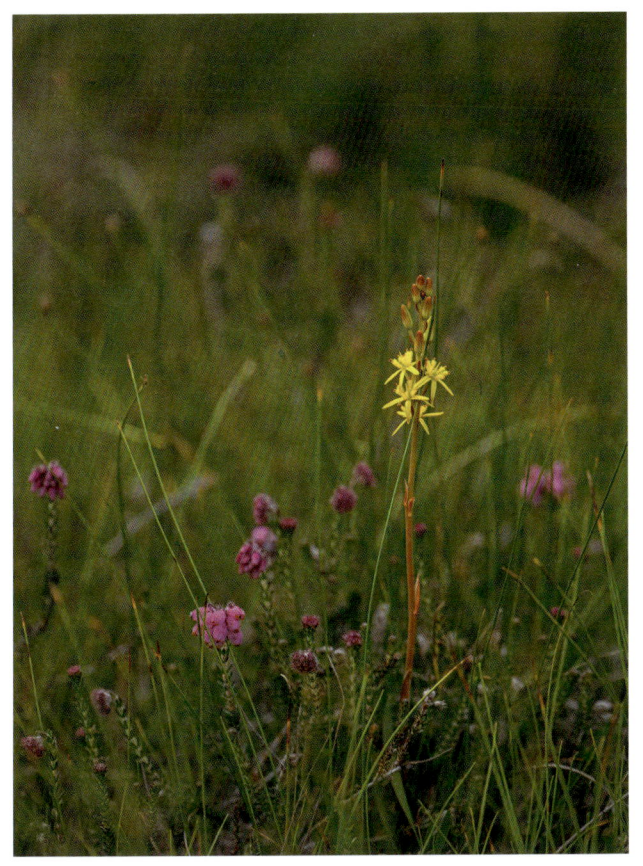

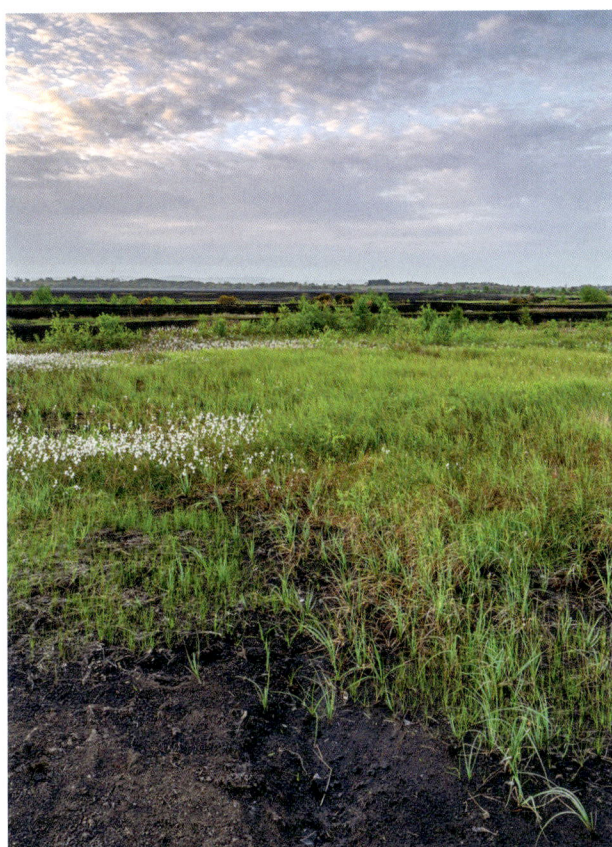

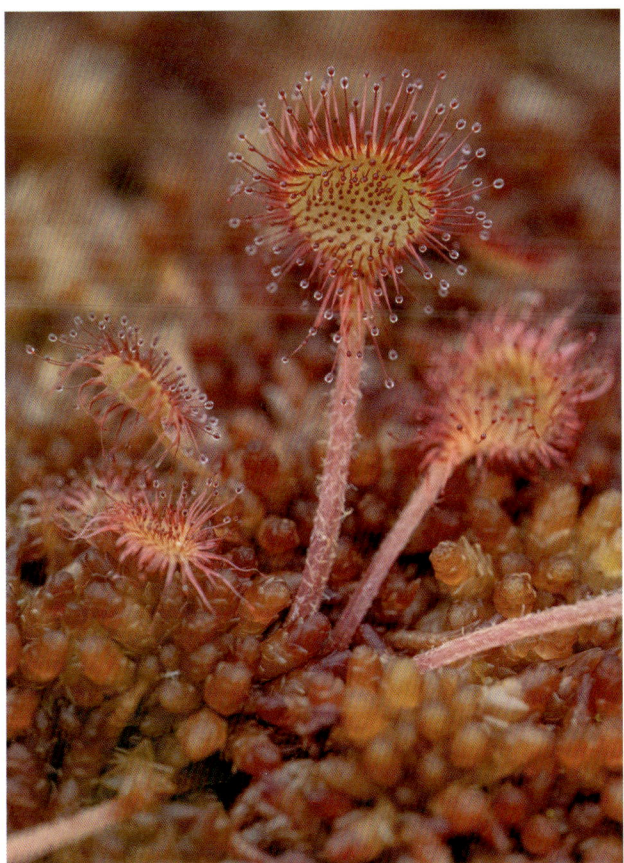

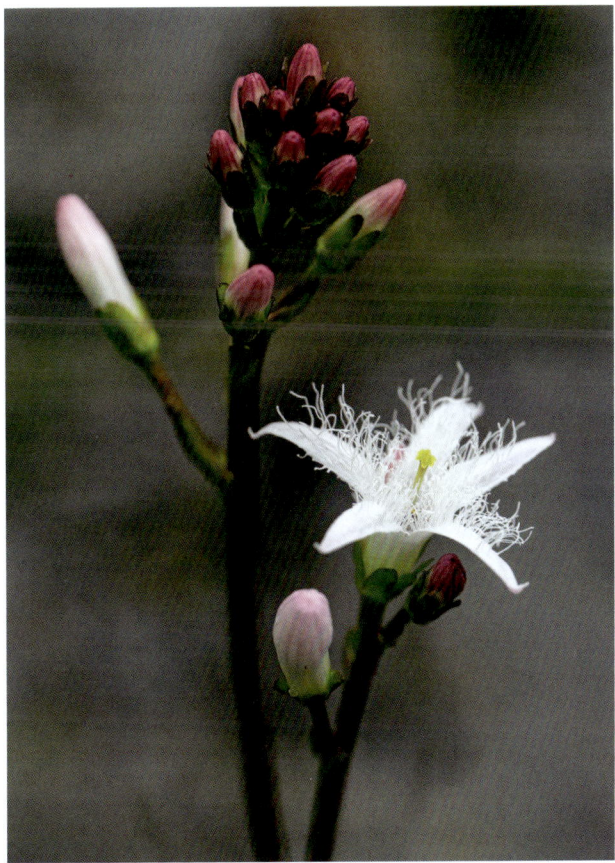

"

BUTTERFLIES AND MOTHS REST ON THESE ISLANDS, THE COMMON LIZARD USES THEM FOR SUNBATHING AND THE IRISH HARE TO DRY ITS FEET

"

of fully intact and growing raised bog surviving, it is hard to imagine the grandeur of this landscape stretching uninterrupted to a distant horizon. Colourful, spongy sphagnum carpets, known as lawns, dotted with fluffy white common cotton grass and white beak sedge dominate the scene. Colonies of tiny sundews and elegant bog asphodels thrive here and there, and a closer look reveals the vibrant cranberry and graceful bog rosemary. In between the sphagnum lawns loom dark hollows and pools, home to bog bean and bladderwort and the realm of the ultimate predator of the bog, the beautiful but deadly raft spider. Somewhat drier hummocks, also built by sphagnum mosses, stand out like small islands and are covered in heather species, mostly ling and cross-leaved heath, hare's-tail cotton grass, deergrass and a variety of lichen and mosses. Butterflies and moths rest on these islands, the common lizard uses them for sunbathing and the Irish hare to dry its feet. This landscape is a potpourri of sparkling green, bright yellow and rich red tones, sprinkled with some white and purple dots in places, the colours accentuated by the film of water that covers everything.

This vibrancy is, however, limited to the very top layer of the bog. Known as the acrotelm, it is less than fifty centimetres deep and consists of a living growth of sphagnum mosses, the plant communities that thrive on it, and recently deceased plants. Below this living layer lies what is known as the catotelm, layer upon layer of dead plant matter in various stages of decomposition that can extend to a depth of twelve metres at the centre of a peat dome. A raised bog is effectively a living carpet of vegetation that floats on a watery concoction of dead and compressed plant particles. Around 90% of a raised bog is pure water, and only the remaining 10% is made of solid material.

Towards the edges of the dome, the peat mantle shrinks to around three metres and then fizzles out. These areas, known as lag zones, are extremely rare today and are very similar to the bog's precursor, the fen. Here, plants can reach the groundwater table, allowing trees like alder and downy birch to form small woodlands and wildflowers to grow among the sedges.

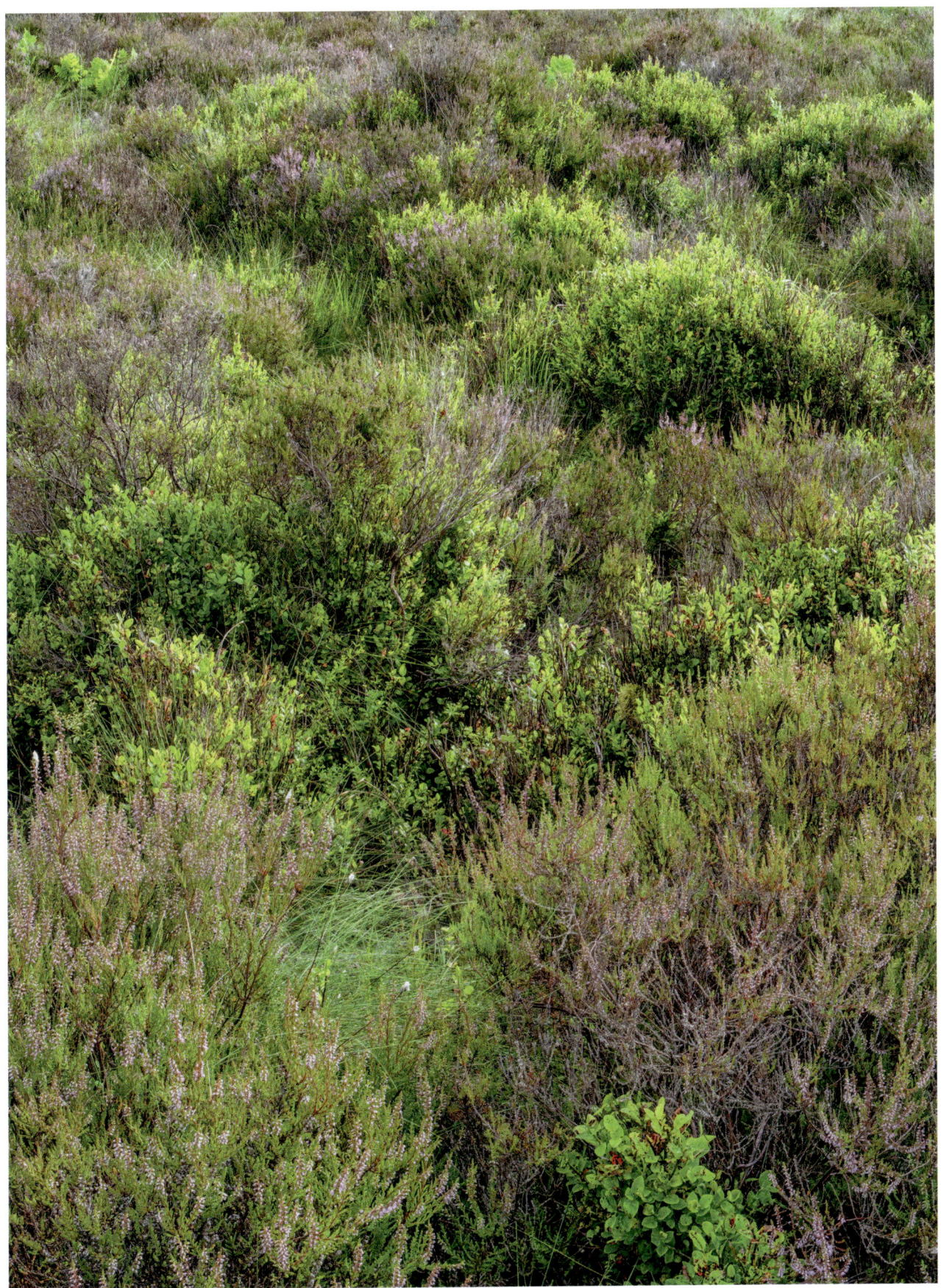

Ling and bilberry

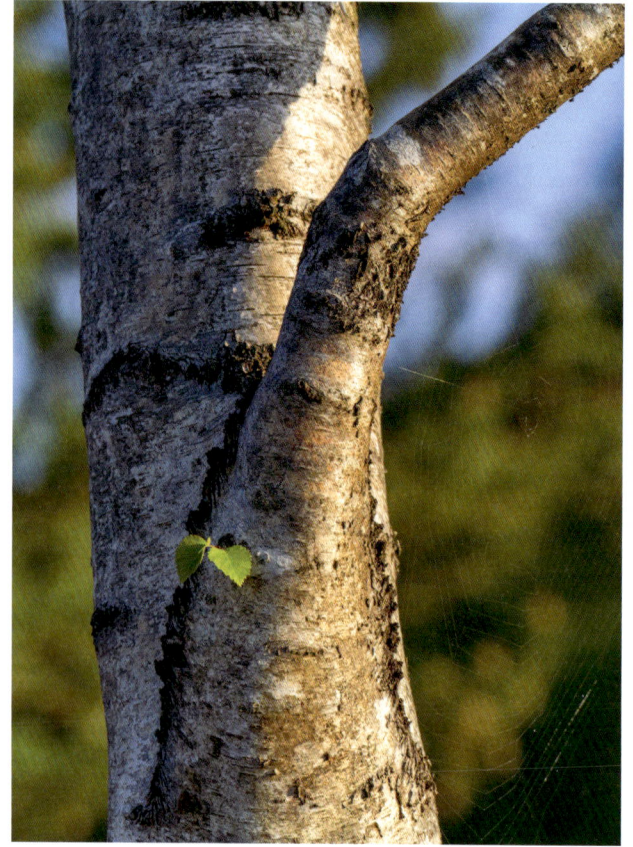

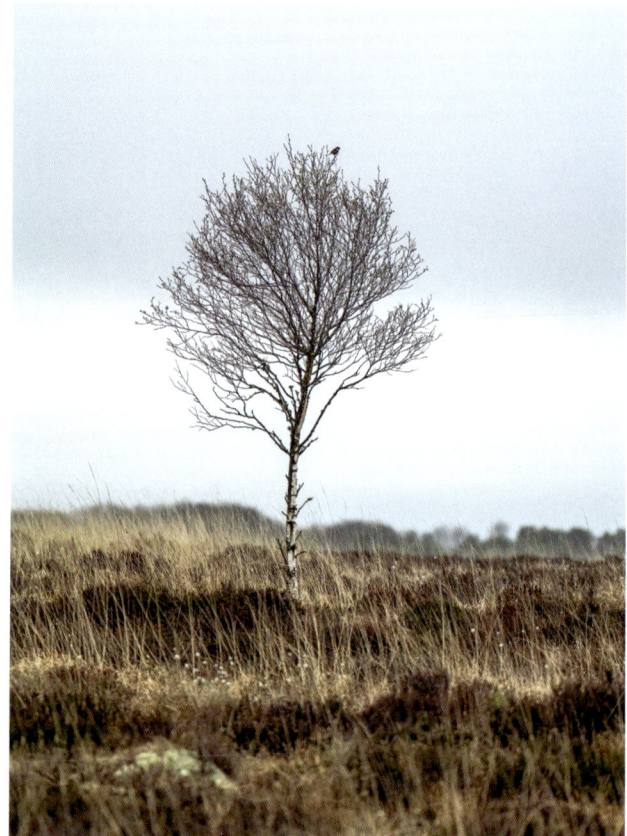

Above left: Silver birch at Killaun Bog, County Offaly
Above right: Silver birch and stonechat at Clara Bog, County Offaly
Opposite top: Smooth newt with emerging butterwort (left) and cranberry plants
Opposite bottom: Monaincha Abbey, County Tipperary

The largest of the midland bog areas is the Bog of Allen, once covering over 1,000 square kilometres and extending over nine counties. In its heyday, the Bog of Allen – which is a collection of many individual peatlands rather than one continuous bog – presented an almost impassable barrier between the east and west of the country. Before and during the Middle Ages, the bog was viewed with a mixture of reverence and trepidation; a mist-laden, seemingly endless expanse of waterlogged, deceiving ground that made life and travel difficult if not impossible. Sometime later, in 1860, the poet Matthew Farrell described the Bog of Allen in a more enchanting way:

Its texture of the richest brown
Set here and there with tufts of down,
Embroid'rd o'er with heath bell trees
And these bespangled o'er with bees
With silky folds of mossy beds,
Like pillows made for angels' heads,
With mirror lakes divided through,
Where countless stars reflect their hue.

Today, little remains of the Bog of Allen's mossy paradise. One remnant is Lodge Bog near Allenwood in County Kildare, whose 10,000-year history saw it transform from a lake to an alkaline (rich) fen to the raised bog of today. This little

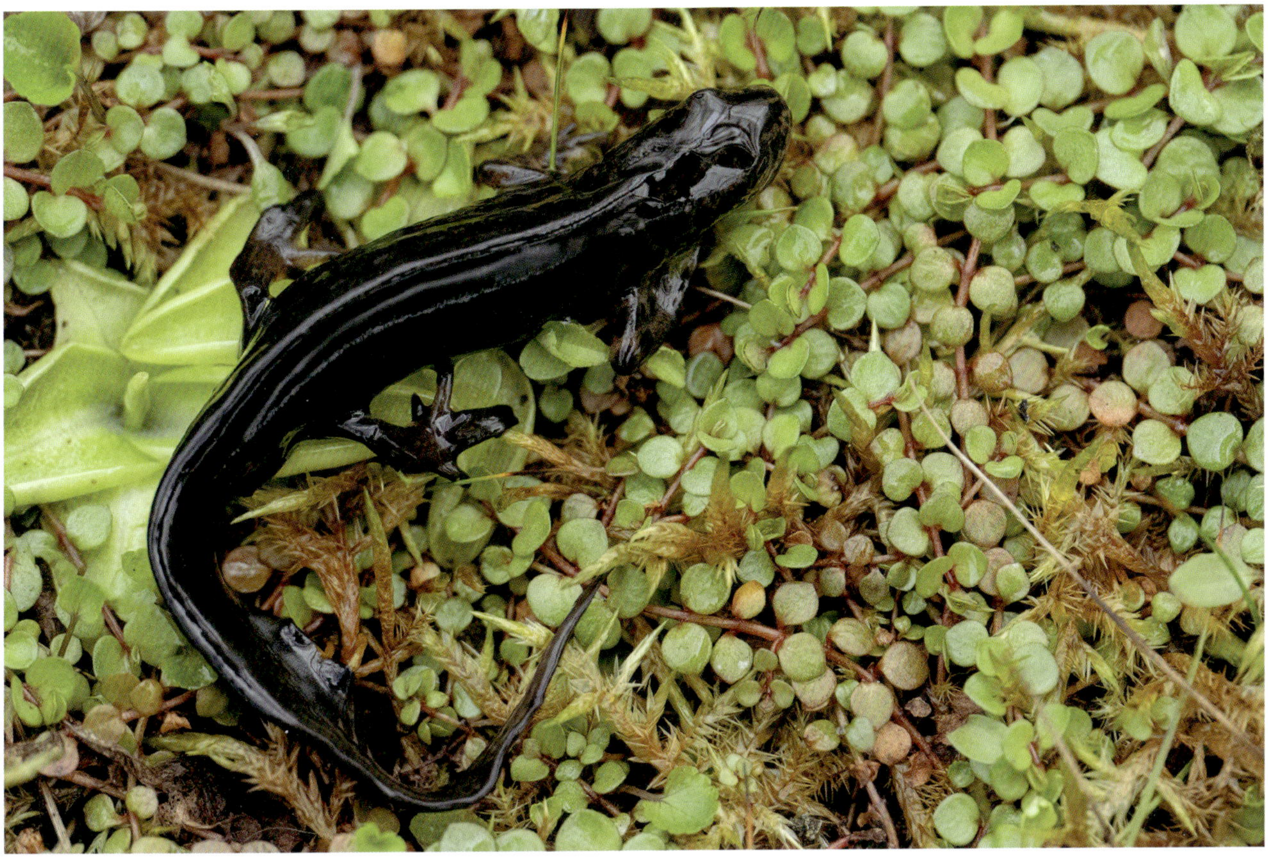

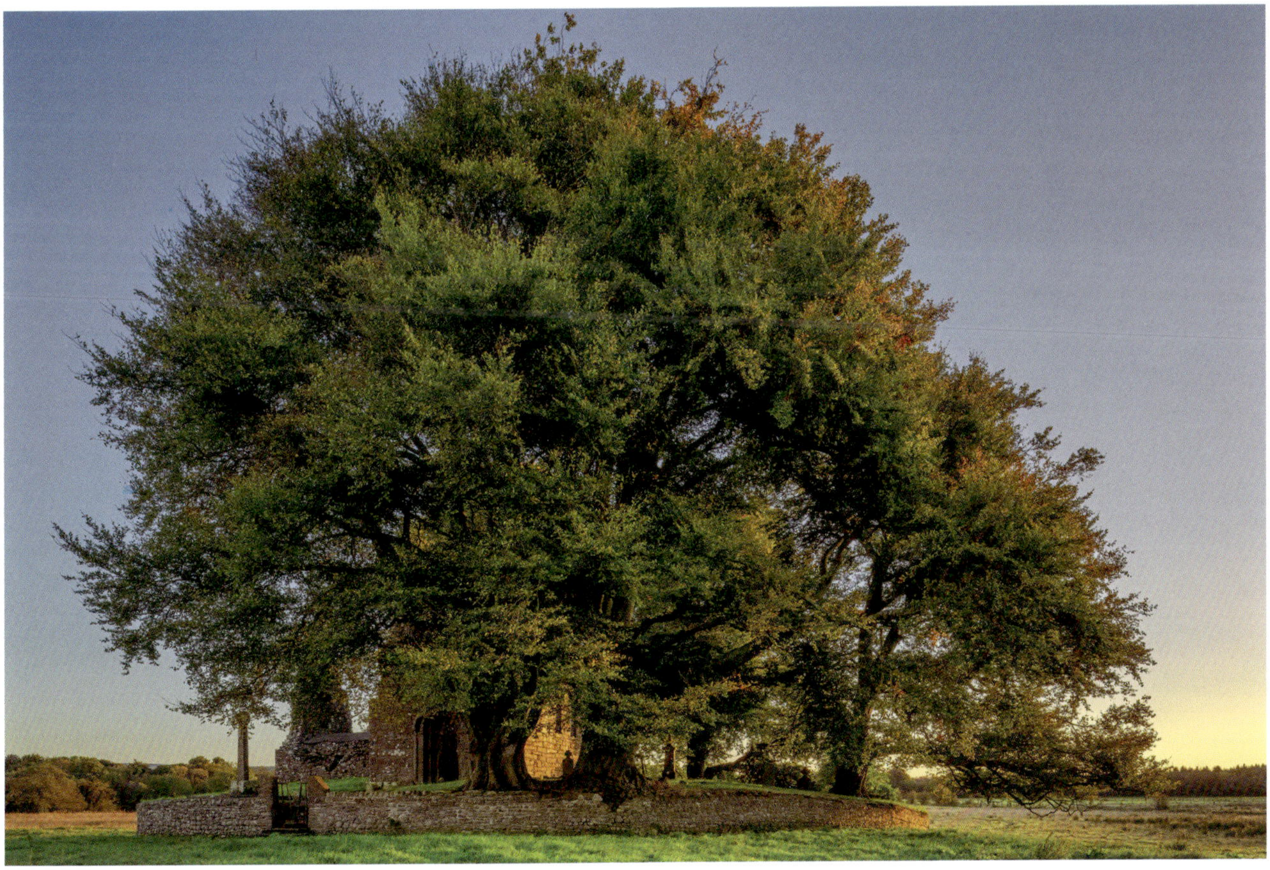

> "
> SOLITUDE WAS IMPORTANT TO THE
> EARLY CHRISTIAN MONKS, AND IT COULDN'T
> GET ANY MORE REMOTE THAN THE MIDDLE
> OF A PEATLAND
> "

peatland is home to some 150 plant varieties, twenty-six bird species, twelve dragonfly species, sixty-one kinds of moth and sixteen of butterfly as well as forty-seven spider species. One of the latter is *Hypsosinga albovittata*, a rare, distinctively marked orb weaver, which was only discovered in 2005 as a new species to Ireland. Lodge Bog is owned and managed by the Irish Peatlands Conservation Council (IPCC), whose headquarters are nearby at the Bog of Allen Nature Centre in Lullymore. This place was once known as Lullymore Island, a small enclave of fertile land in the midst of the peaty ocean that was the Bog of Allen. The island could only be reached by a narrow causeway, which was part of a wider network of highways that ran along the ridges of eskers, back then the only safe passages if you wanted to avoid the treachery of the bog. Naturally this remote spot became a Christian

monastery in the 5th century, its foundation allegedly ordered by St. Patrick himself, whose footprint can be seen to this day engraved into a limestone boulder.

Of course, Lullymore was not the only monastic settlement of this kind. Seeking solitude to study and pray without distraction was important to the early Christian monks, and it couldn't get any more remote than the middle of a peatland. One of the best-preserved peatland refuges is located a little over ninety kilometres to the southwest of Lullymore near Roscrea in County Tipperary. Monaincha, which translates to 'the boggy isle', is a 12th-century monastery built in the middle of a bog lake on the Bog of Monela, an outlier of the Bog of Allen. Today the monastery sits slightly elevated by a manmade foundation in the middle of a large pasture, but it still exudes an air of tranquillity and remoteness.

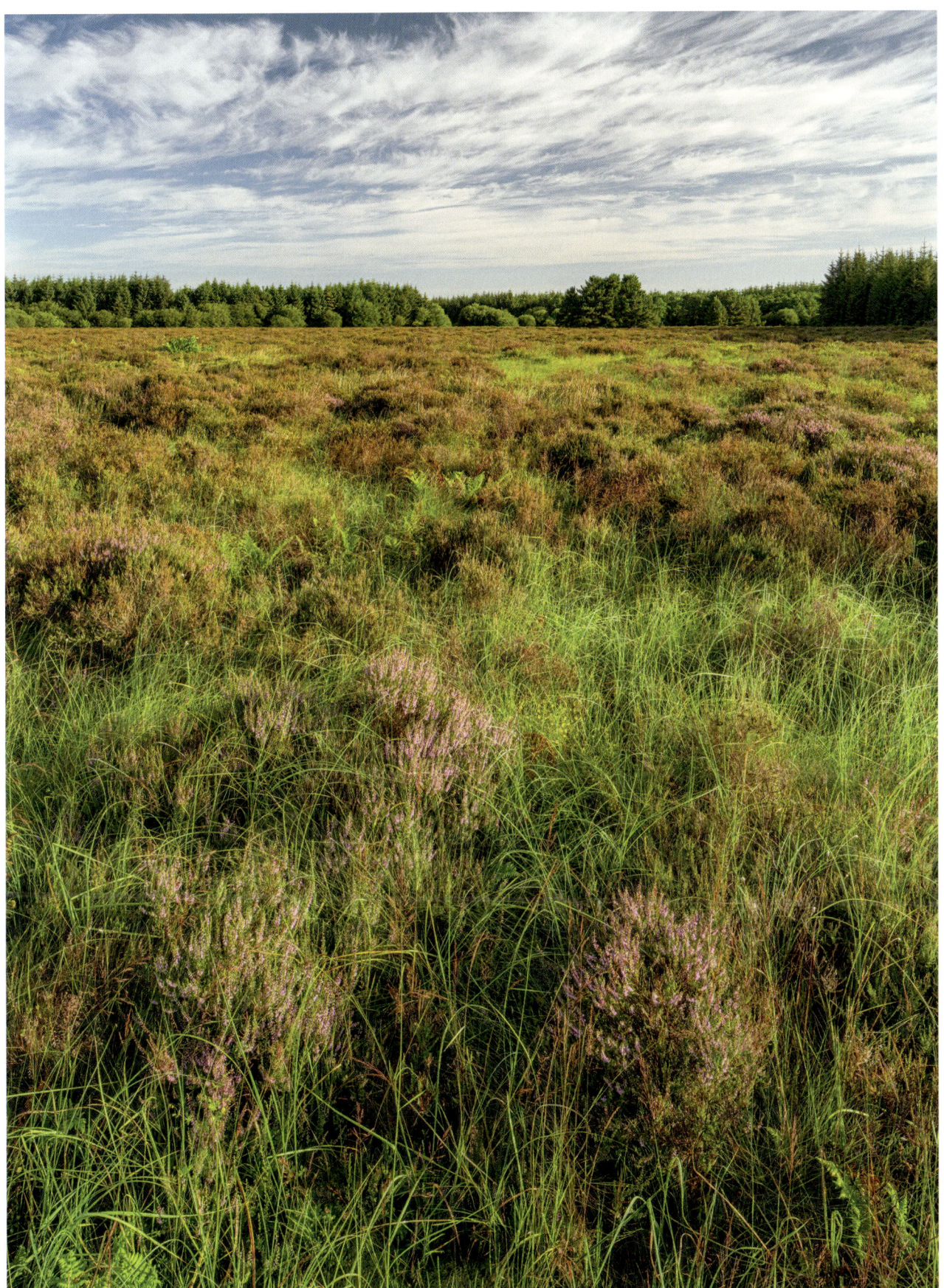

Griston Bog, County Limerick

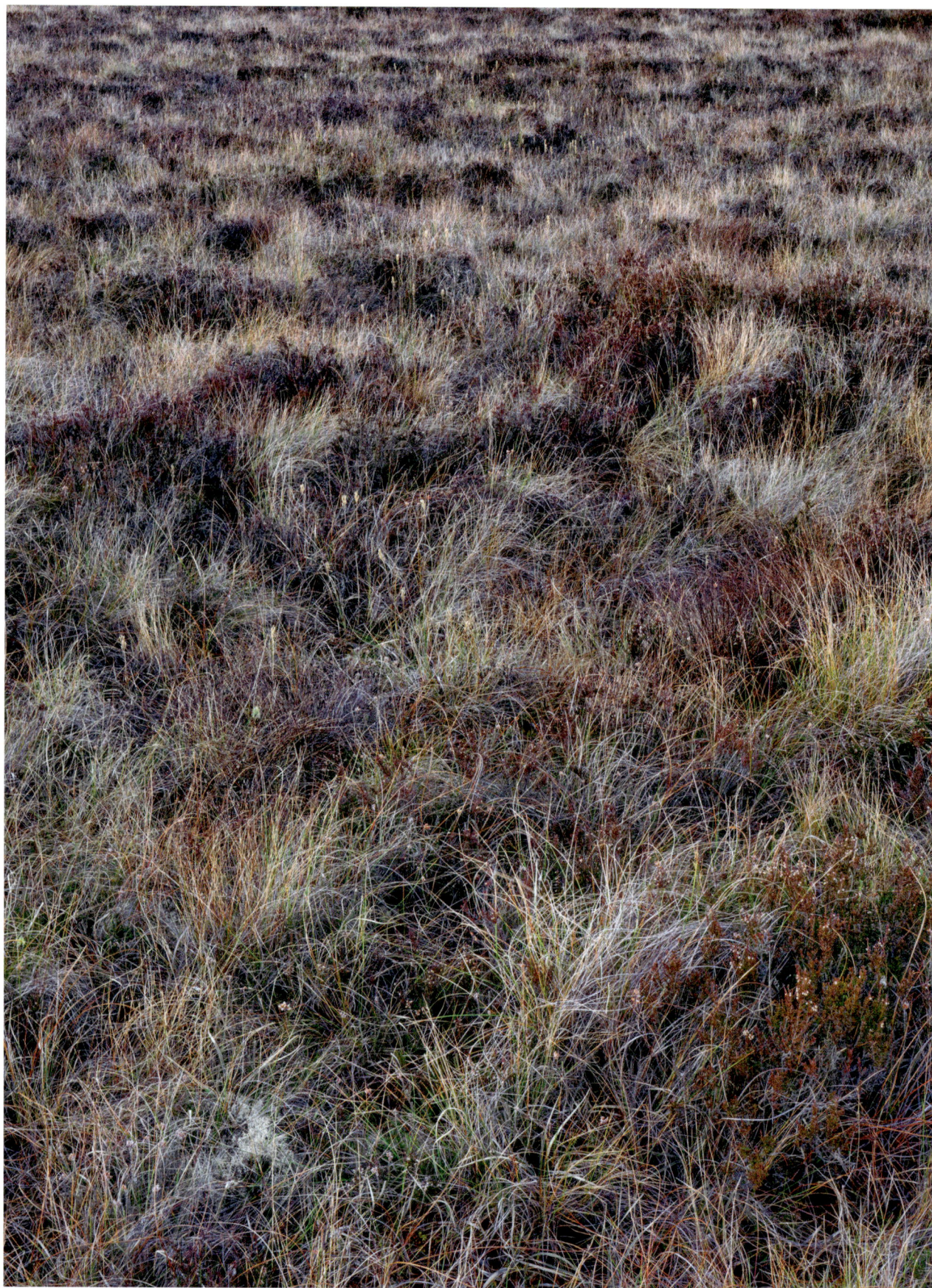

Winter at Clara Bog, County Offaly

"

A STAGGERING 156 MOTH SPECIES WERE RECORDED AT LULLYMORE WEST, AS WELL AS TWENTY-ONE BUTTERFLY SPECIES, INCLUDING THE RARE MARSH FRITILLARY

"

Just west of the Bog of Allen Nature Centre lies Lullymore West, also owned and managed by the IPCC. While this area is no longer a raised bog, it has somewhat recovered from decades of peat harvesting and today features a mixture of scrub and wet grassland, a haven for moths and butterflies. In 2006, a staggering 156 moth species were recorded here, as well as twenty-one butterfly species, including the rare marsh fritillary.

One of the best-preserved and largest of the remaining raised bog areas can be found in County Offaly. Clara Bog only escaped the fate of its neighbouring peatlands, which were all harvested by Bord na Móna, because of intense public protests in 1983 that led to the site being purchased by the Irish state and declared a national nature reserve in 1987. In addition to the typical features and inhabitants of a raised bog, Clara Bog is home to an unusual soak system, a small area fed by nutrient-rich ground-water that supports a fen-like plant community, including a small woodland of downy birch, in the middle of the acid peat dome. Another specialty of Clara Bog is the narrow cruet-moss (*Tetraplodon angustatus*), a beautiful bryophyte whose tightly packed shoots form light green tufts among the sphagnum species.

Clara Bog covers a bit over eight square kilometres, only half of which is considered to be pristine. This is reflective of the bigger picture. It is estimated that once over 3,000 square kilometres of the Irish midlands were covered in raised bog. In 1974, only 650 square kilometres remained, and by 1985, it was a meagre 200. Today, only around 25% of the remaining raised bogs are considered to be in a conservation-worthy condition, and even less, around 10%, are recognised as fully intact. They are small islands of memories from a distant past, and dreams for a wilder future.

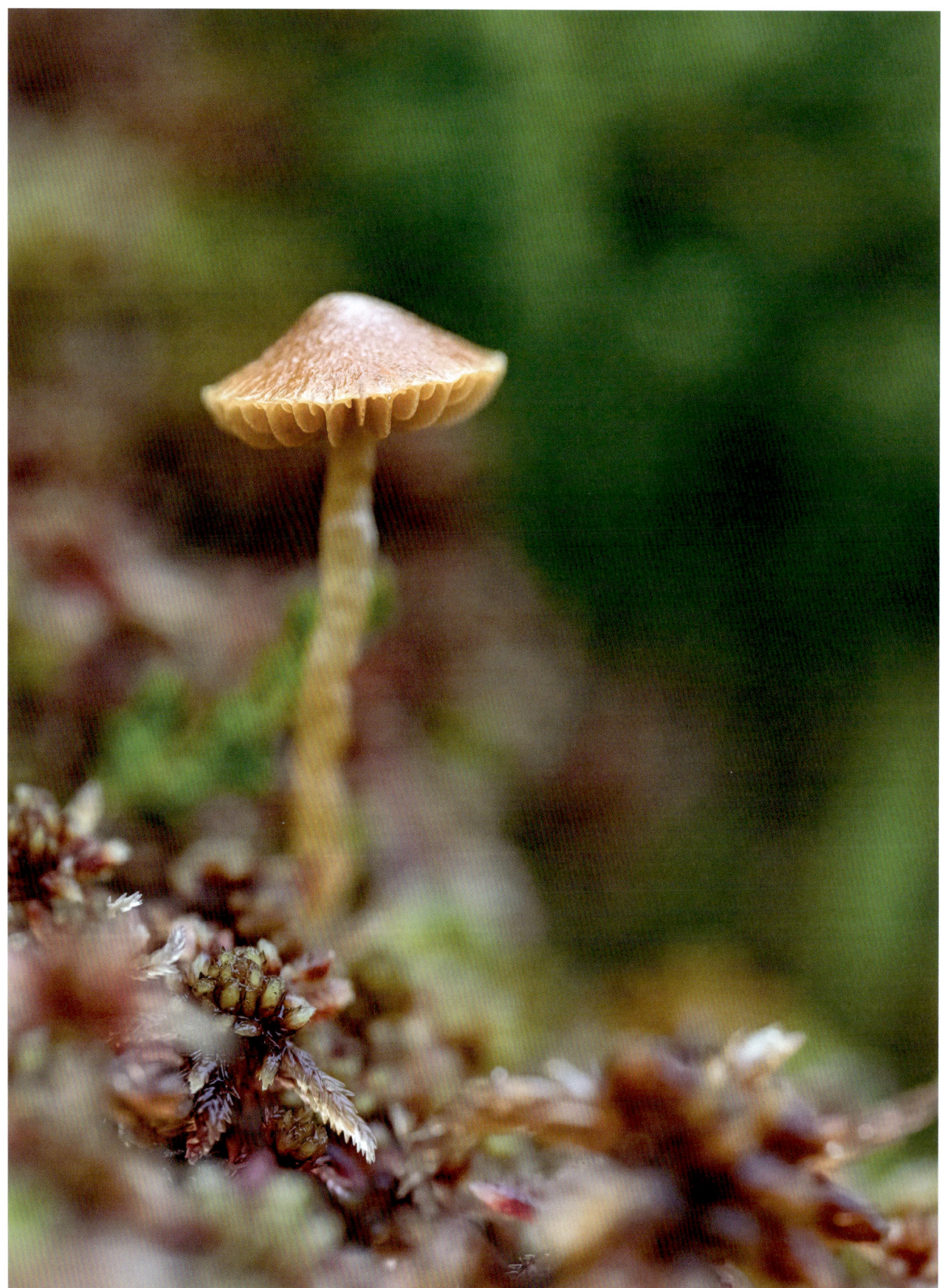

Mycena megaspore growing among sphagnum mosses

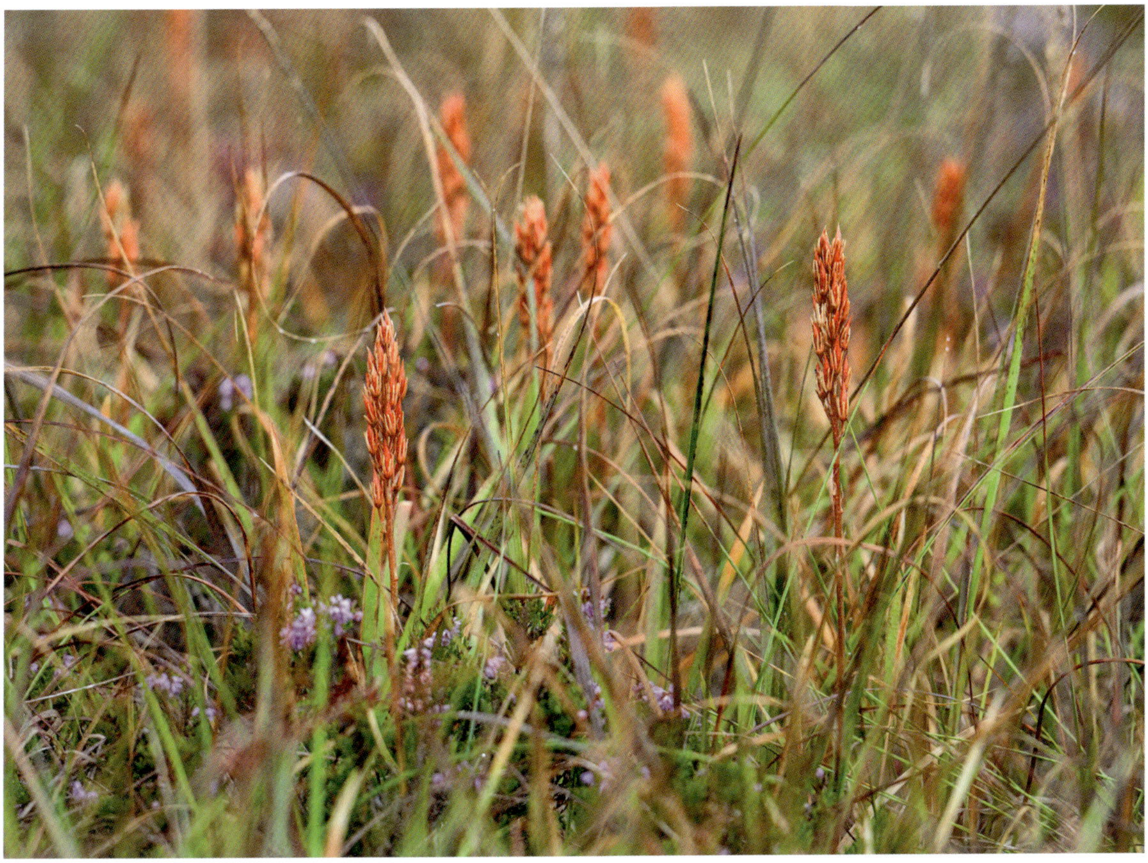

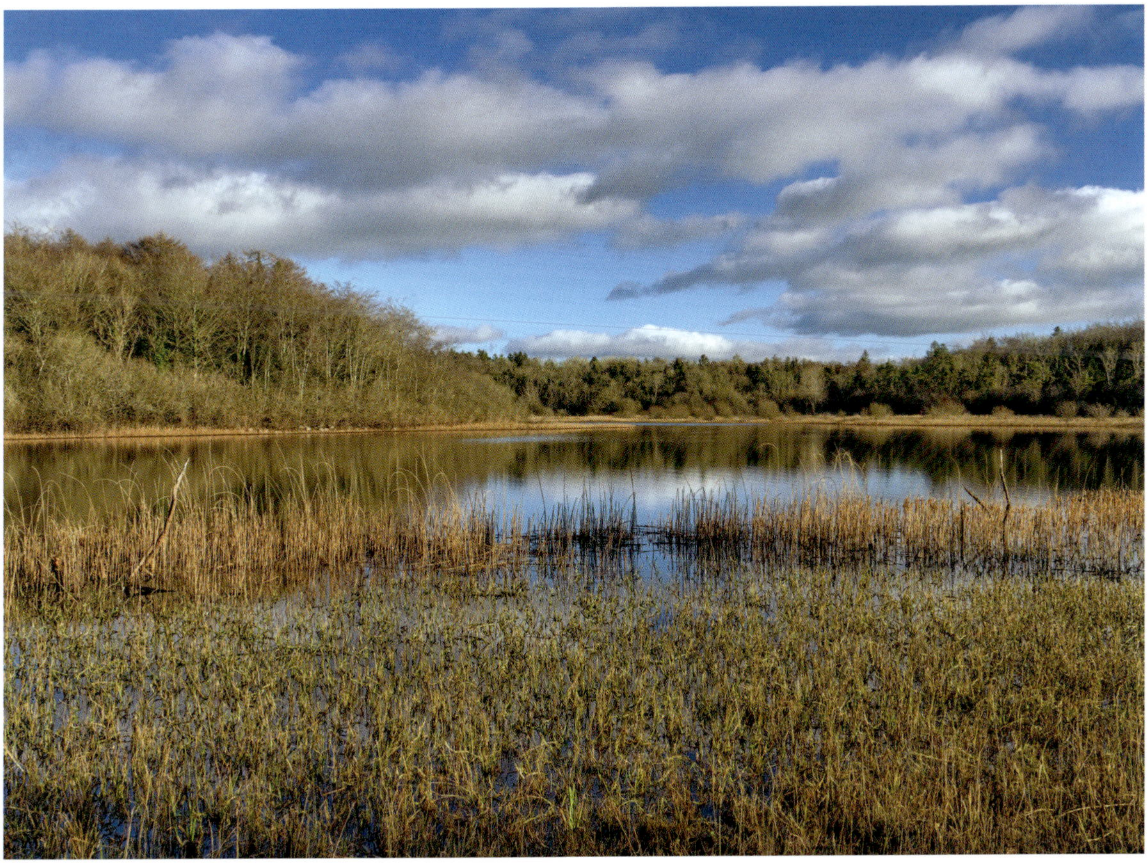

Top: Bog asphodel in autumn • **Above:** Fen landscape, Dromore Nature Reserve, County Clare

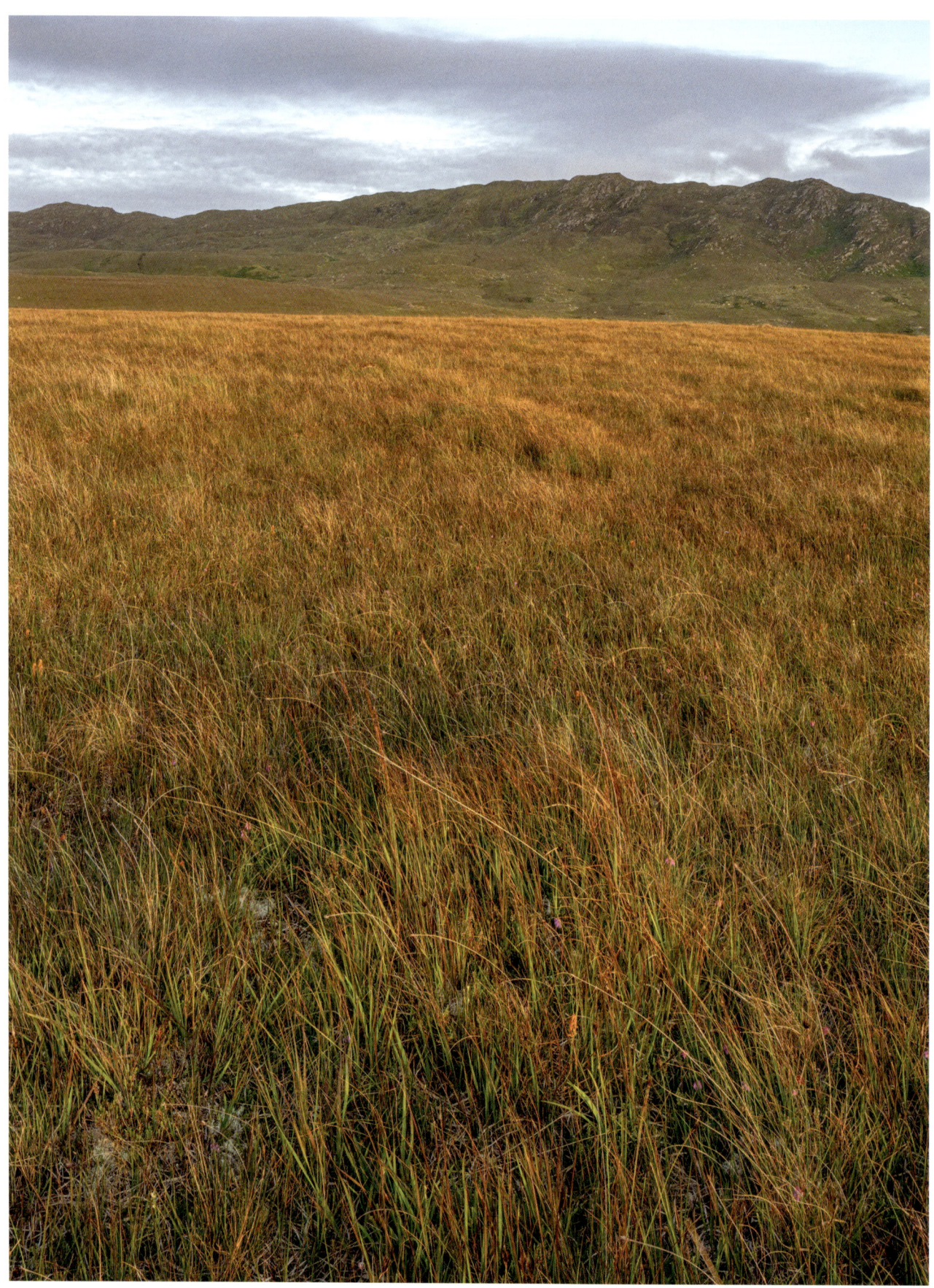

Summer evening at Claggan Mountain, County Mayo

Chapter Three

BLANKET BOG, HEATH AND MARSH

The landscape of the west of Ireland couldn't be more different to the gentle and domesticated countryside of the midlands. In the southwest, some of Ireland's highest mountain ranges, namely the Macgillycuddy's Reeks and the Brandon Range, stretch their sandstone peaks towards the sky. Further north, Connemara and Joyce's Country feature rugged plains and the unmistakable quartzite domes of the 12 Bens and Maumturk Mountains. County Mayo displays the epitaph of emptiness with the vast open landscapes of Ballycroy and Bangor overshadowed by the distant Nephin Beg mountain range. In the northwest, County Donegal portraits itself with the sweeping granite hills of Derryveagh rising between spacious valleys. What all these landscapes, despite their varying underlying geology, have in common is a habitat that is as beautiful as it is desolate, the blanket bog, covering the land from the coastal fringes up into the mountains.

On a global scale, blanket bog is a comparably rare habitat, restricted mainly to the boreal regions of the northern hemisphere and the western seaboards of Ireland and Scotland. Ireland holds 8% of the world's blanket bogs, most of them located on the plains and mountains of its western counties.

The origin of this landscape, like that of the peatlands of the midlands, lies in a distant past. The circumstances that led to the development of blanket bogs were, however, very much different to those that gave rise to the fens and raised bogs. After the glaciers of the Ice Age had retreated some 10,000 years ago, a tundra landscape established itself and over time developed into habitats of vast grasslands and forests that covered the plains and extended into the mountains. In the west, this landscape was built onto a shallow layer of soil. Beneath that soil lay a foundation made of acidic bedrock. In the west and northwest, these rocks were granite, quartz and rhyolite; in the southwest, red sandstone. As a consequence, the soil overlaying the bedrock was also slightly acidic, which turned out to be the first puzzle piece needed for the growth of peat. As long as the relatively dry boreal climate – characterised by long, cold winters and short, warm summers – prevailed, peat only managed to develop in areas of poor drainage, where water could accumulate on the surface. In these pockets, the waterlogged ground in combination with the acidic conditions created by the bedrock

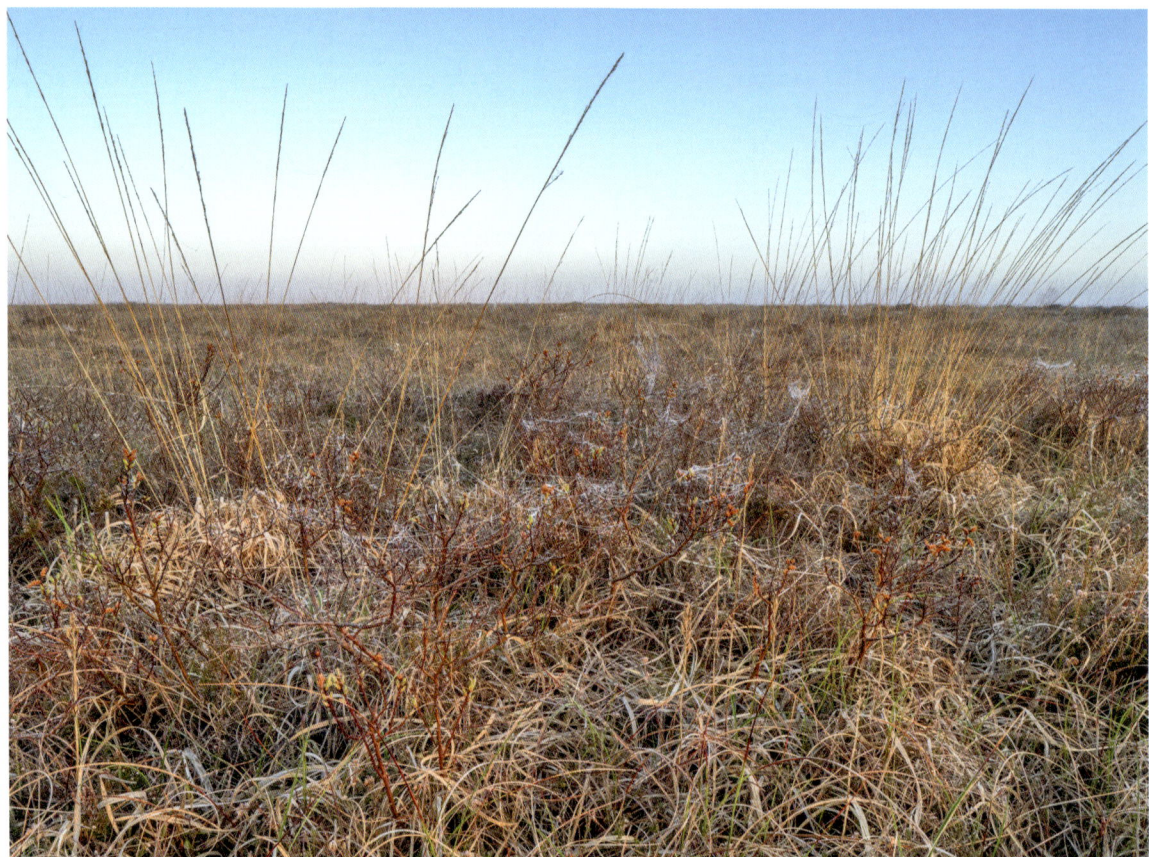

Above: Winter at Tullaher Bog, County Clare

Opposite top left: Foggy sunrise in the bog
Opposite top right: Black Valley, County Kerry
Opposite bottom left: Sphagnum
Opposite bottom right: Friars Glen, County Kerry

prevented dead plant material from completely decomposing. This material accumulated and over time and under its own weight was compressed into the first peat layers. In the post-glacial landscape, these pockets appeared as areas of marsh and heath as well as swamp woodland as early as 9,000 years ago.

These embryonic blanket bogs were confined to their wet cradles until the continental climate regime was replaced by today's Atlantic oceanic climate. While this new regime, with its increased precipitation and year-round mild temperatures, influenced all of Ireland and played a major role in the formation of the raised bogs of the midlands, its greatest effect was felt along the western seaboard. Here, the vast mountain chains caught much of the incoming moisture and forced it on the landscape as rain, drizzle, fog, hail and snow. This constant watery onslaught caused minerals to be washed from the soil and accumulate in what is known as the iron pan, an impermeable layer in the ground that caused the soil above it to become permanently waterlogged. The wet climate and water-saturated ground were the other pieces of the puzzle the blanket bog needed to start its reign. Blanket bog needs precipitation of 1,250 millimetres a year spread out over at least 250 days to thrive, and the new conditions provided just that. Invigorated by the constant water supply, the young peatlands left their birthplace and spread

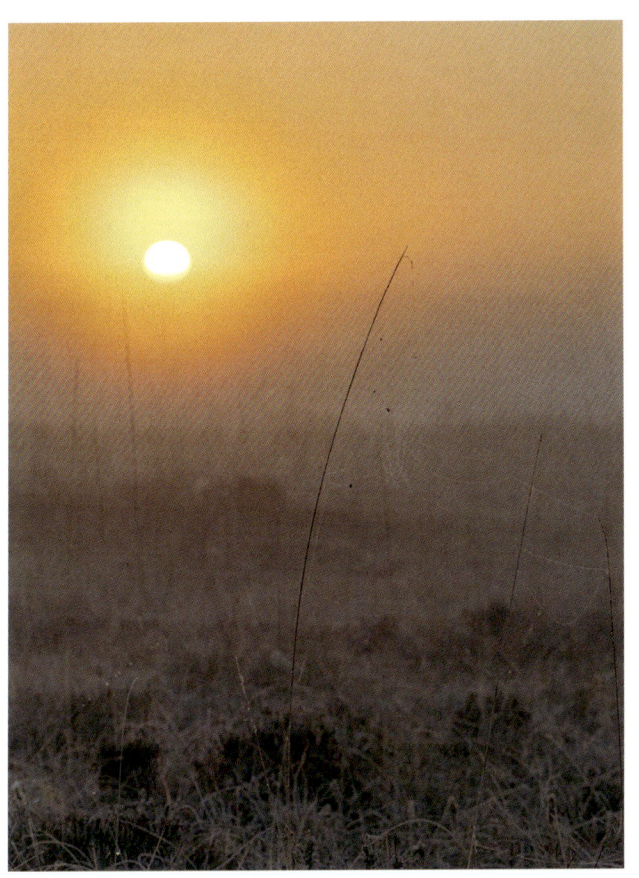

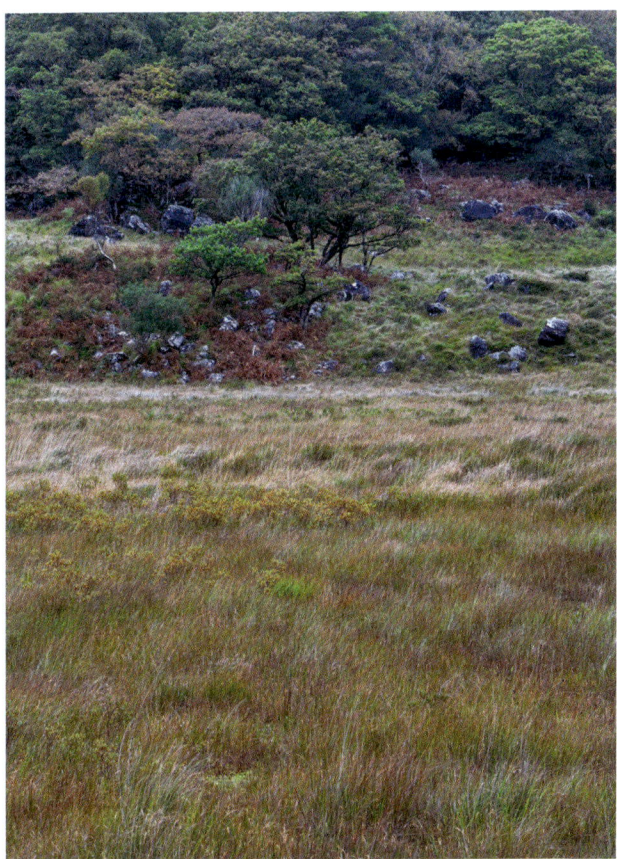

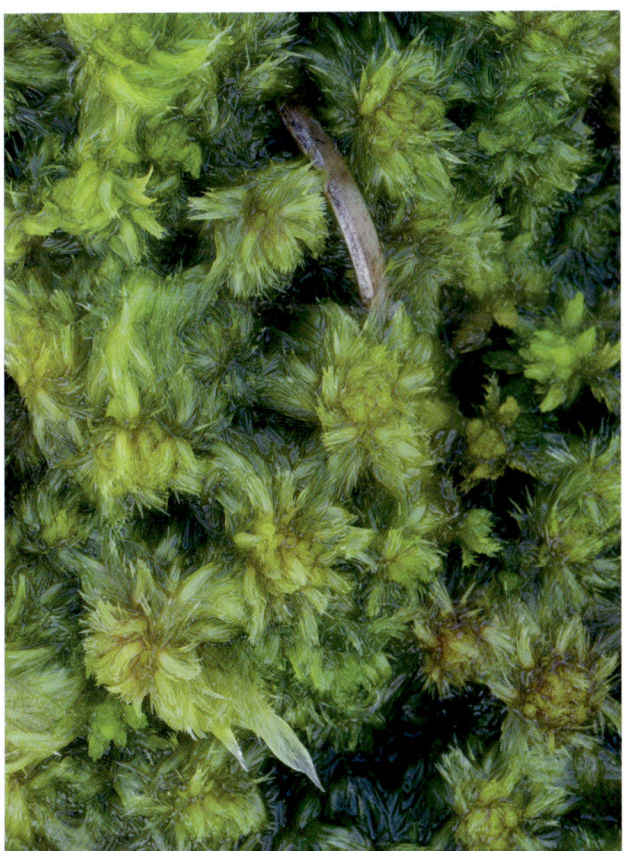

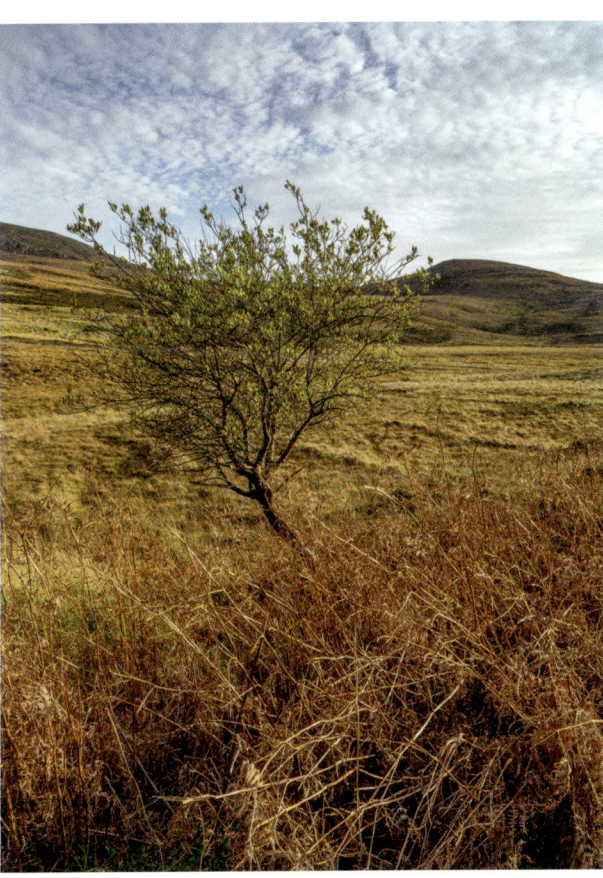

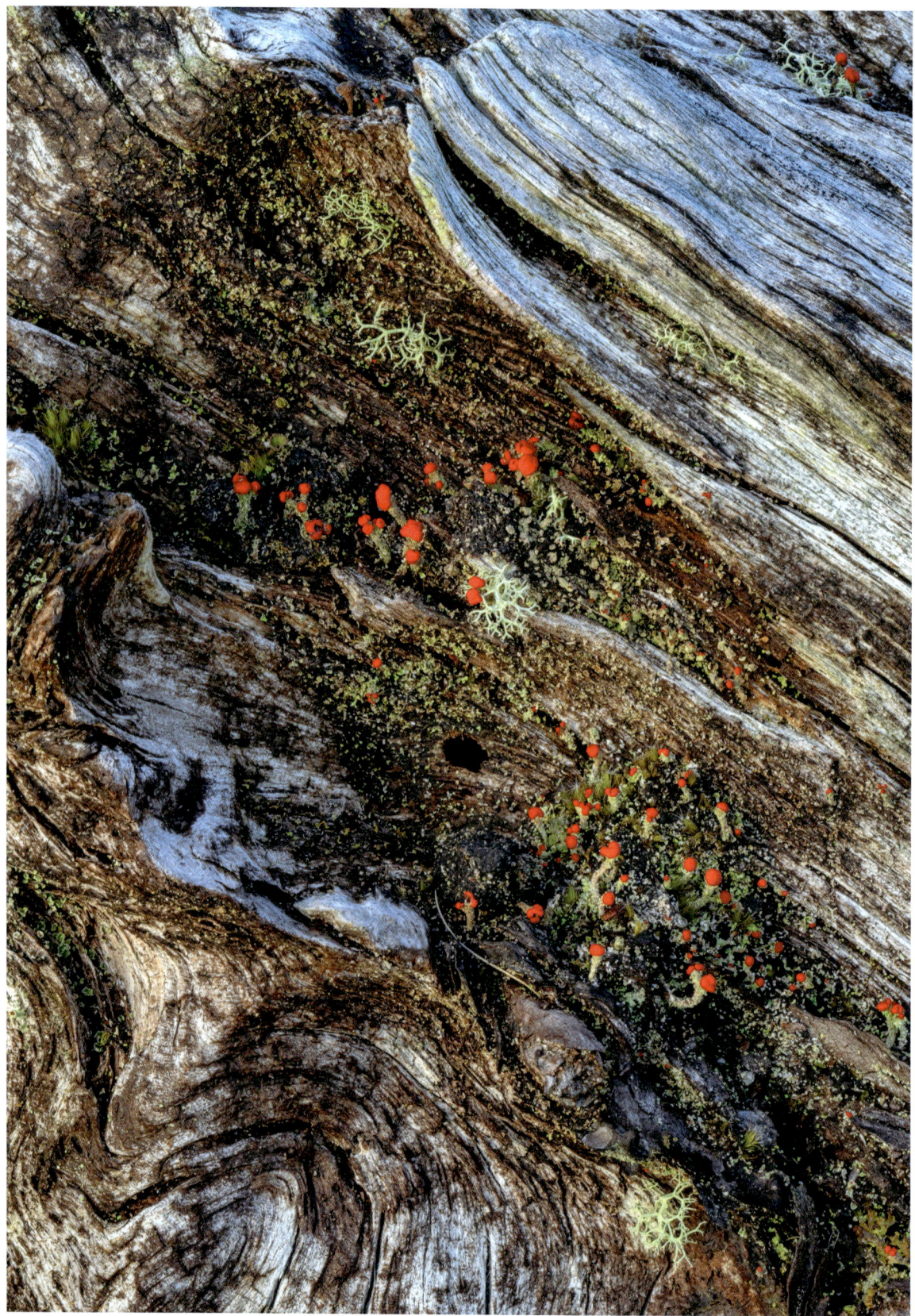

Lichen on bog oak

"

THE WHITE, FLUFFY FLOWERS OF THE COMMON COTTON GRASS HAVE STUFFED PILLOWS AND BEEN SPUN INTO THREADS TO MAKE CLOTHES

"

over the landscape, engulfing anything in their way.

As the ground became permanently waterlogged and peat layers began to build up, plants that couldn't adapt to the new living conditions disappeared and were replaced by the characteristic blanket bog flora. Sphagnum mosses, which play such a vital role in the raised bogs of the midlands, are considerably less prominent and only play a minor role in blanket bog formation. Instead, grasses, rushes and sedges are the dominating plant families on blanket bogs and are a good indicator of the average water saturation of any given area. The tuft-forming purple moor grass, for example, can tolerate permanently waterlogged ground, while deer sedge, also known as deer grass, prefers somewhat drier areas where it forms firm tussocks.

The differences between grasses, rushes and sedges are subtle and reveal themselves only on closer inspection. All three plant families feature narrow, elongated leaves, but while the leaves of grasses are mostly soft and flexible, rushes and sedges are more rigid. Many rush and sedge species also show a characteristic V-shaped cross section and feature sharp edges. The flowers of grasses are elaborate, complex, and very much vary from species to species. Rushes and sedges, by contrast, keep it simple and mostly feature only one inconspicuous flower cluster on top of the flower-carrying stem. This stem also differs between the three families. The flowering stem of grasses is usually round and hollow and shows distinct joint-like nodes. Rushes also have hollow stems but lack the nodes and are more solid. The sedges' stems lack the hollow centre and are often not perfectly round but display a subtle triangular cross section.

The most eye-catching of the lot are the Eriophorum species, the cotton grasses that despite their common name belong to the sedge family. The most widespread of the four species present in Ireland is the common cotton grass, also known as bog cotton. The plant got its name from the appearance of its flower heads, which very much resemble the ones of the cotton plant and have also seen a similar use. The white, fluffy flowers of the common cotton grass have stuffed pillows and been spun into threads to make clothes; both, however, only as a measure of last resort when down feathers and wool were not available. Bog cotton clothes made in Scotland were exhibited at the Great Exhibition in 1851 in London, but because of its lack of tensile strength, the regular use of common cotton grass in the textile industry never took off.

Common cotton grass is a perfect example of how well adapted sedges are to their environment. The plant grows in the wettest parts of peatlands, even right inside bog pools. It can do so because it has developed a snorkelling device: interconnected air spaces known as aerenchyma that connect the upper parts of the plant to the roots, which can reach up to sixty centimetres into the ground. These air canals ensure the gas exchange between the upper and lower part of the plant, particularly the oxygenation of the root. The latter is important so the cells in the root can burn glucose, which powers the ability of the roots to absorb water and nutrients from the ground.

Hare's-tail cotton grass, identified by its smaller, less fluffy and more defined seed heads, lacks this snorkelling feature and can therefore only be found on drier peatland areas. Less widespread than the common and hare's-tail cotton grass is the slender cotton grass, which was only discovered in 1966 and is known exclusively from a few sites in Connemara, two sites in County Kerry, and one each in County Mayo and County Westmeath, where it thrives at the edges of pools and on the vegetation matt of quaking bogs. Somewhat more widespread but still one of the rarer sedges is the broad-leaved cotton grass, a plant very similar to the common cotton grass. The two species only differ slightly in the appearance of their leaves. As the name suggests, the broad-leaved cotton grass has broader and flatter leaves, while common cotton grass has very narrow (the *angustifolium* in its Latin name stands for narrow) and V-shaped leaves.

Grasses, sedges and rushes don't just contribute to the growth of peatlands by becoming one of its components. The species' main adaptations to life on peatlands, their deep-reaching fibrous roots, stabilise the soft and fragile peat substrate and so effectively ensure the integrity of the landscape. These roots, in addition to anchoring the plant in place and collecting nutrients, also perform another task. Many species have adapted parts of their root structure into rhizomes and stolons, which allow them to spread and colonise new territory without being dependent on seeds, and the resulting underground network further stabilises the peat.

Above ground, the plants have also adapted to the challenging environment. Their leaves are narrow, and some have developed a waxy surface to keep water loss through evaporation to a minimum. In a waterlogged place like a blanket bog, this doesn't seem to make much sense, but water intake through the root systems in the oxygen-deprived and acidic conditions of peatlands is difficult for the plants, while the typically cool and windy Irish weather causes a speedy evaporation.

—————•—————

Opposite top left: Common cotton grass
Opposite top right: Bog asphodel
Opposite bottom left: Common reed
Opposite bottom right: Bog bean

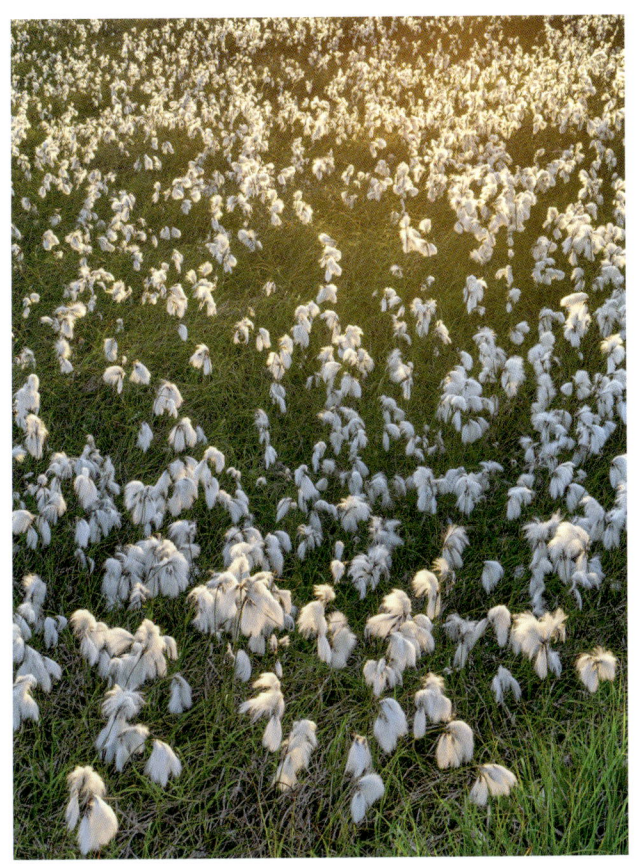

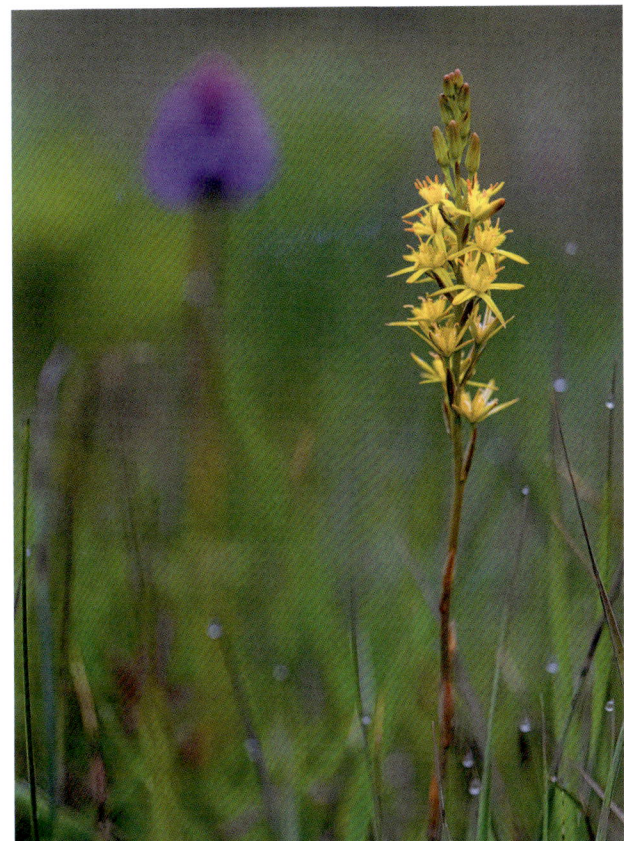

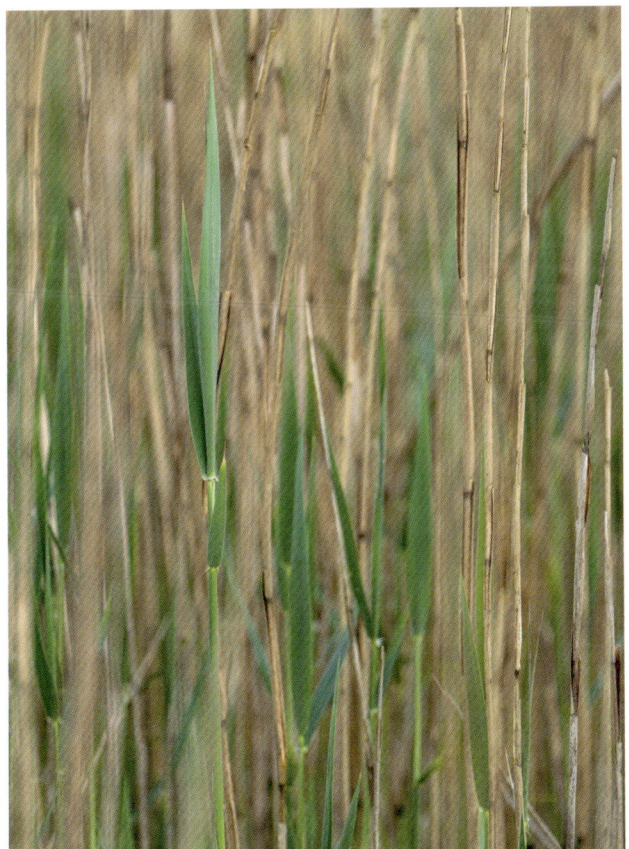

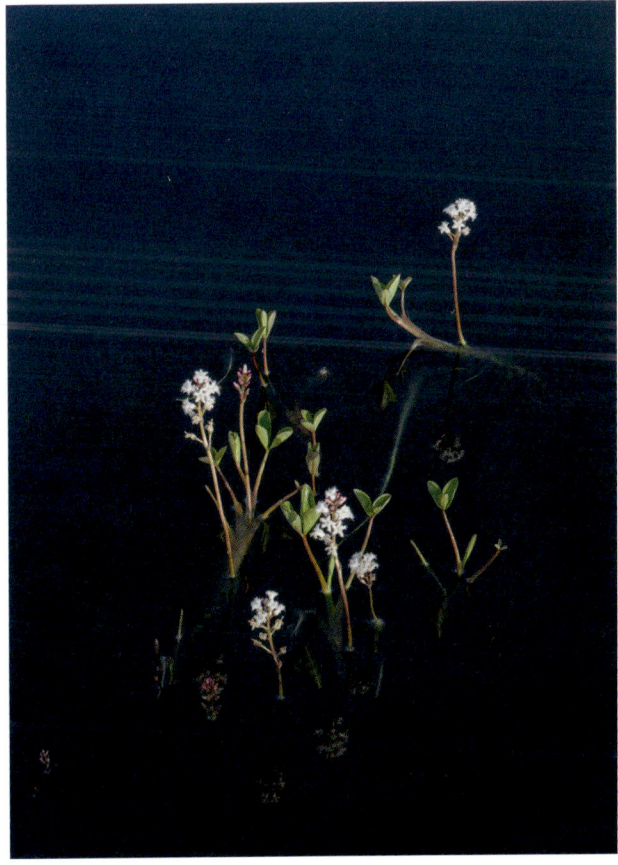

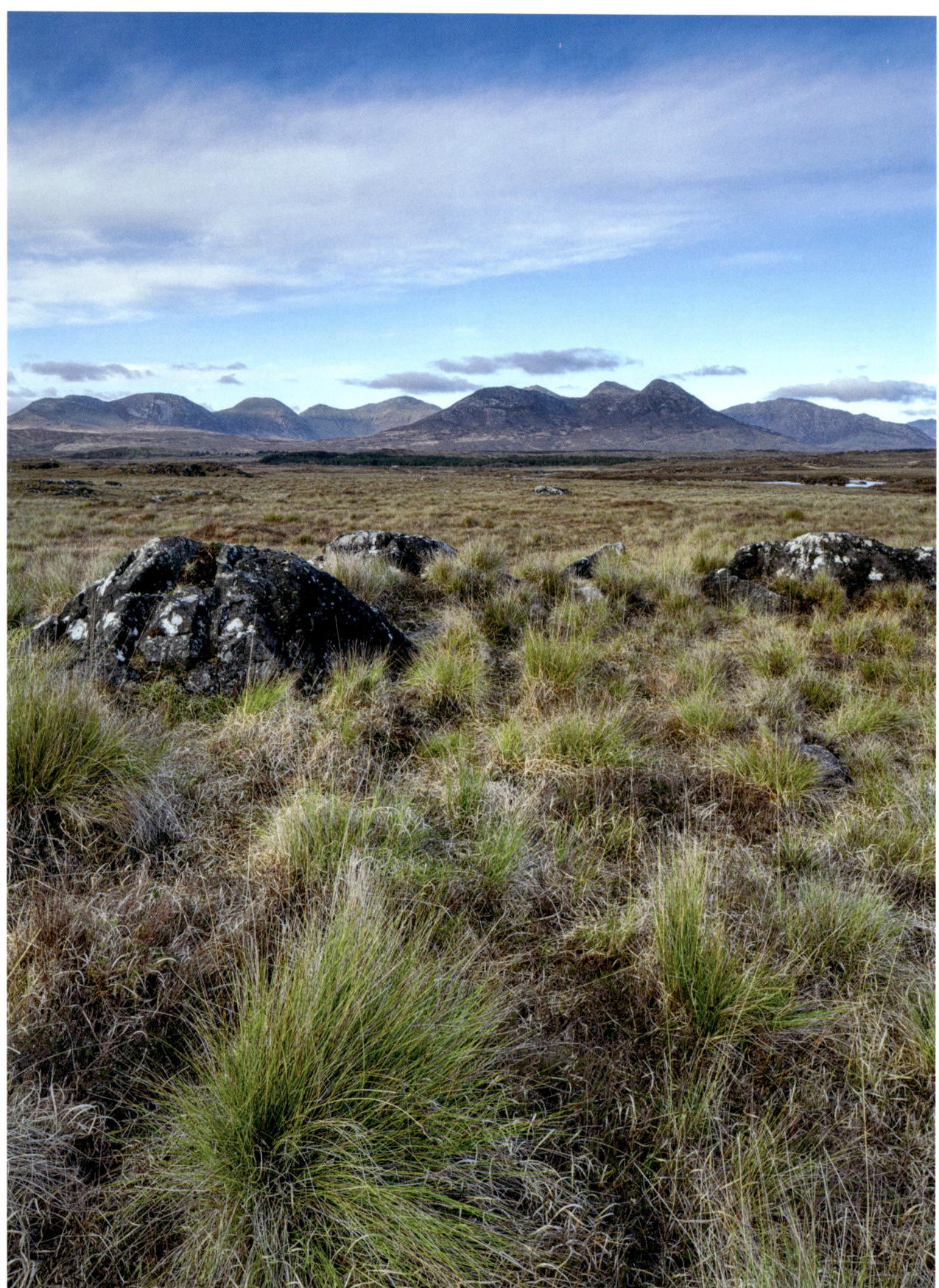

Roundstone Bog, Connemara

"

THE CÉIDE FIELDS ARE ONE OF THE OLDEST KNOWN FARMING LANDSCAPES IN THE WORLD

„

Standing at the shore in northern County Mayo, looking east towards the Nephin Beg range of mountains, it becomes very obvious how the blanket bog got its name. The apparently homogenous surface extends from the water's edge uninterrupted into the distance, accentuating the contours in the landscape like a blanket spread over a sleeping giant, stretching up into the mountains and only escaping the view on the peaks of the faraway hills.

The domination of the blanket bog happened fast, but the process was not entirely straightforward. While it took only a few centuries for the peat to engulf the landscape and oust the previous vegetation, slight changes in weather patterns and short-lived drier periods allowed trees, mostly birch and pine, to periodically recolonise the peat. The remains of this so-called pine-woodland phase are evident in tree trunks that reappear where peat layers have been removed, either through human extraction or natural erosion. Despite these interludes, the blanket bog had a firm grip on the landscape. Nowhere else is this more evident today than in northern County Mayo on a hillside overlooking the Atlantic Ocean.

In the 1930s, local schoolteacher Patrick Caulfield was cutting his turf supply for the winter when his slean hit something hard. Patrick started digging and discovered the remains of an old dry-stone wall. Further investigation by archaeologists revealed not only a whole network of boundary walls but also houses and burial structures, all dating back some 6,000 years. The Céide Fields – the 'fields on the flat-topped hill', as the area became known – are one of the oldest known farming landscapes in the world. Over a period of several hundred years, a community of late Neolithic and early Bronze Age farmers grew their crops and raised their livestock in this now blanket bog-covered, bleak and windswept place. Several millennia ago, however, the hillside must have looked very different. Growing crops in the west of Ireland has likely never been easy, but back then stands of pine and birch provided protection for small, enclosed fields and their thin layer of fertile soil. In the beginning, this soil provided well for the small community, but the population grew, and the yearly weather patterns began to change.

Exploration of the area has shown that all farming activity did stop rather suddenly, which leads to the conclusion that a substantial event must have taken place on the flat-topped hillside. A likely scenario is that the changing climate and subsequent advance of the blanket bog made the land unsuitable for farming. Another theory suggests that the farming activities themselves were a trigger for the growth of the peatland. During the end of the Neolithic and in the early Bronze Age, farming tools and techniques were becoming ever more sophisticated, allowing for a more intensive use of the land. This resulted in the clearcutting

"

LOWLAND BLANKET BOG IS MORE SPECIES RICH AS IT RECEIVES NUTRIENTS BLOWN IN FROM THE ATLANTIC IN THE FORM OF SEA SPRAY

"

of woodlands to make space for the creation of larger fields. Once a critical number of trees had been removed, there was nothing left to stabilise the ground, and the increasing precipitation started to leach nutrients and erode and wash away the soil. The formation of the iron pan did the rest, waterlogging the ground, and so a blank canvas was created, ready for the blanket bog to move in. If human activity was a deciding factor in the rise of the blanket bog or if farming only sped the process along can't, however, be answered for sure.

Blanket bog comes in two variations, lowland blanket bog and upland blanket bog. They tend to mingle with heaths, and together these ecosystems form an intricate patchwork in which the individual habitats are difficult to distinguish.

Lowland blanket bog is present from the coast up to roughly 150 metres above sea level, where upland blanket bog takes over, and some sources go further and classify upland blanket bogs found 300 metres or more above sea level as mountain blanket bogs. The main difference between the two types is in the prevailing flora. Lowland blanket bog is more species rich than upland blanket bog, as it receives some additional nutrients blown in from the Atlantic in the form of sea spray. The dominating plants here are grasses, rushes and sedges, of which black bog rush, purple moor grass, deergrass and white beak sedge are the most common. A variety of well-adapted wildflowers also thrive in the lowlands, including heath milkwort, tormentil, lousewort, bog asphodel and the carnivorous sundews and butterworts. Pools, ponds and lakes occurring here host a rich flora that includes bladderworts, water lobelia and pipewort. Small shrubs can also be present, mostly bog myrtle and heathers. The latter, however, are more widespread in and characteristic of the uplands.

In upland blanket bogs, heathers, mostly ling and cross-leaved heath, are abundant and can dominate the vegetation. A variety of grasses, rushes and sedges like purple moor grass, deergrass and cotton grasses are also present, but black bog rush – widespread in the lowlands, where the nutrients it needs are in abundance – becomes rarer with increasing elevation and distance from the coast. Two indicator species for upland blanket bog are bilberry and crowberry, the latter being an essentially alpine species most at home at heights of 300 metres and more above sea level. Bilberry also occurs in lower elevations and can be found in woodlands and raised bog, but it very rarely appears in lowland blanket bog.

Another obvious, but (unless the bog has been harvested or eroded) invisible clue to the type of blanket bog is the depth of the peat. Lowland blanket bogs can have peat layers up to eight metres thick, while peat in the uplands is on average one to two metres shallower. There are exceptions to that rule, however, and deeper peat accumulations can occur in upland areas with little or no steep gradients.

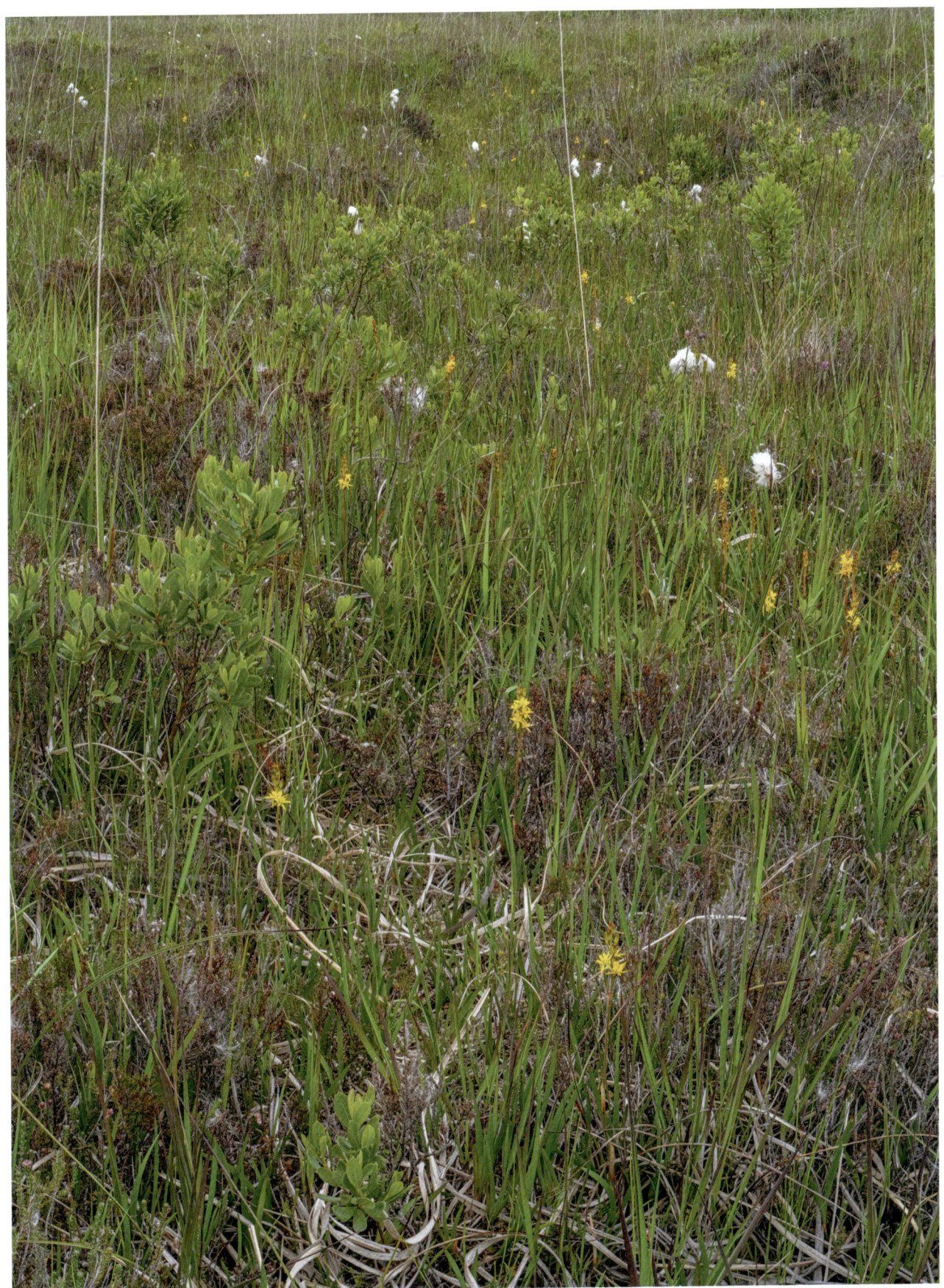

Blanket bog flora

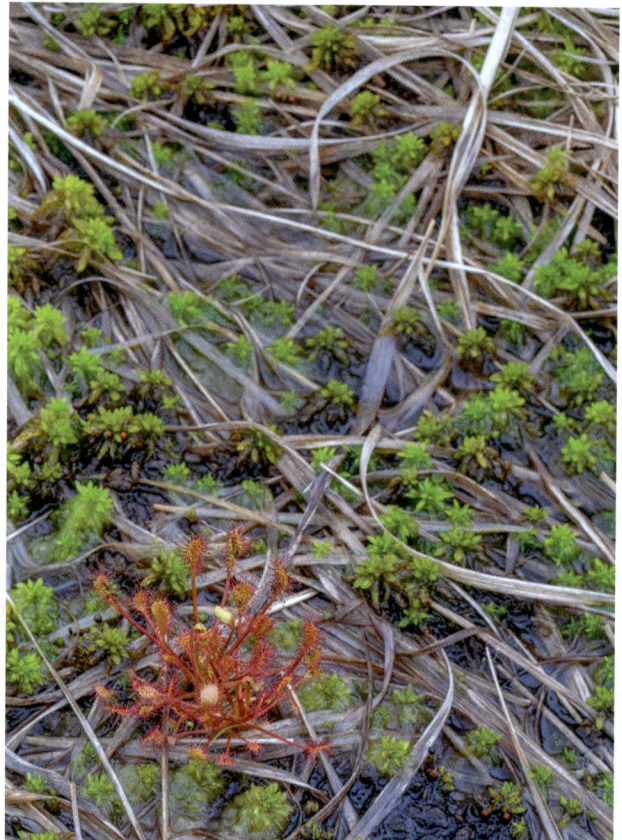

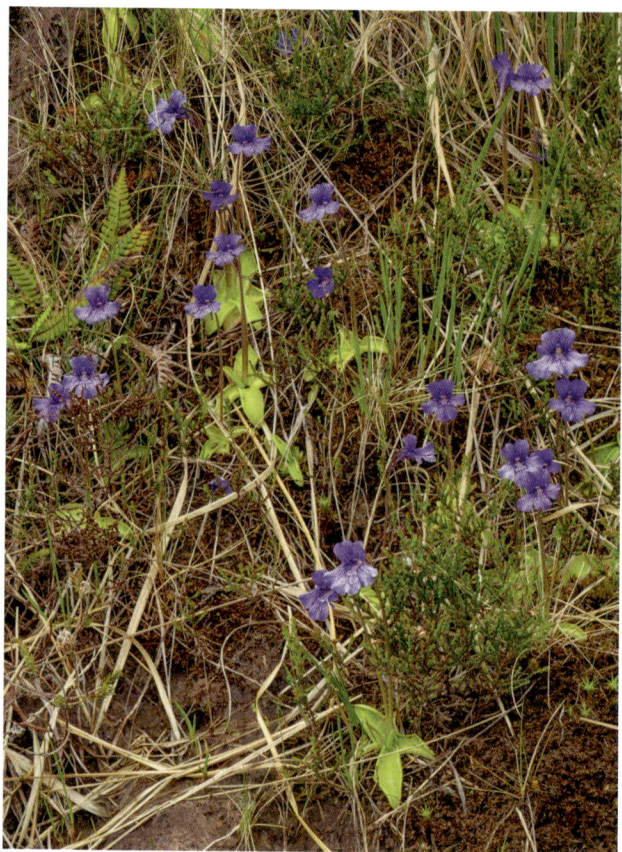

The depth of the peat in heath, a habitat which regularly appears as a part of lowland and upland blanket bog, is considerably shallower than that of any blanket bog. Heath develops in areas that are reasonably well drained and subsequently less waterlogged. The peat layer in heath can be anything from a few millimetres to fifty centimetres at the most, and in some heaths, it might even be completely absent.

Heath comes in three main types: dry calcareous heath, dry silicious heath and wet heath. Dry calcareous heath can be found in well-drained limestone areas and is therefore very rare along the west coast. It is nevertheless an interesting habitat. Dry calcareous heath develops when a thin, base-rich soil layer leaches its nutrients, making the environment unsuitable for more demanding plants. Instead, a flora not dissimilar to that of peatlands moves in. Ling, bell heather, purple moor grass and tormentil flourish alongside calcium-loving, so-called calcicolous plants like burnet rose and kidney vetch, and shrubs like hazel, juniper and blackthorn are all part of the characteristic dry calcareous heath vegetation.

Above left: Oblong-leaved sundew
Above right: Large flowered butterwort
Opposite top: Old cottage at Roundstone Bog, Connemara
Opposite bottom: Mountain blanket bog at Wicklow Mountains National Park

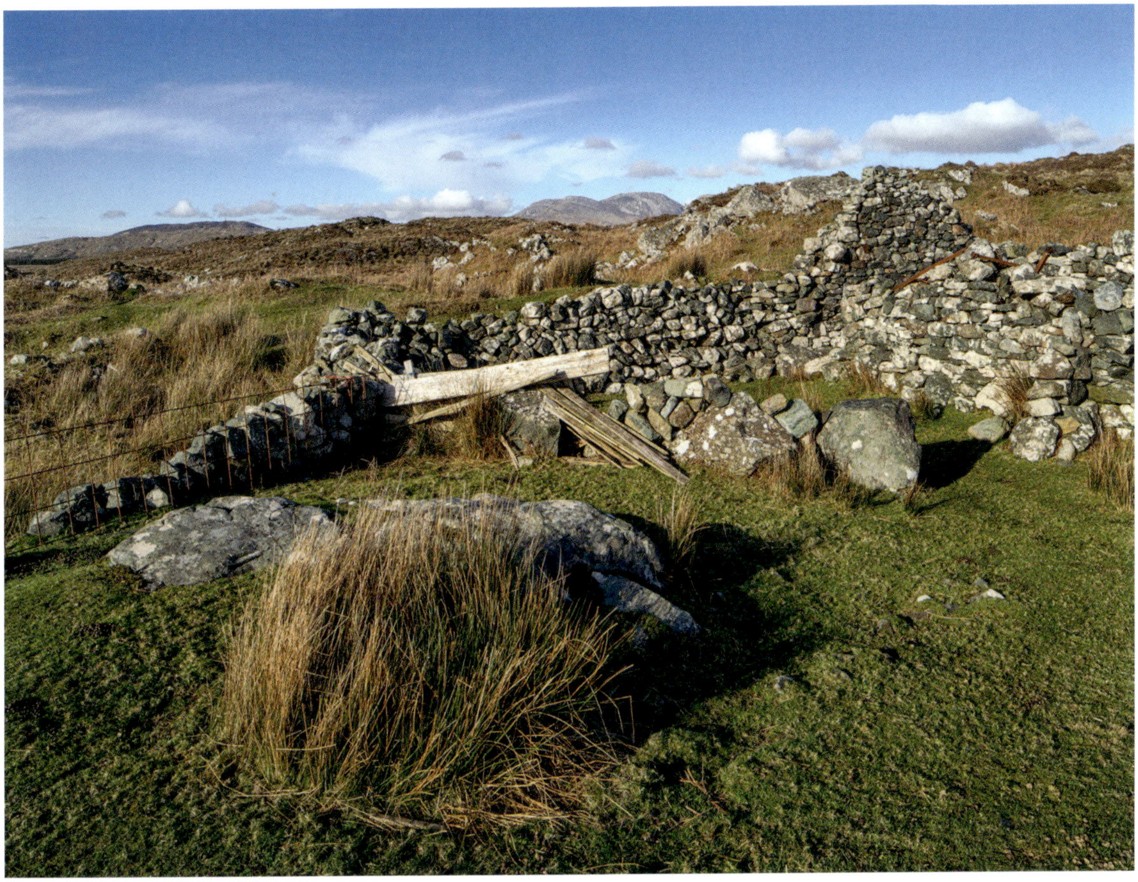

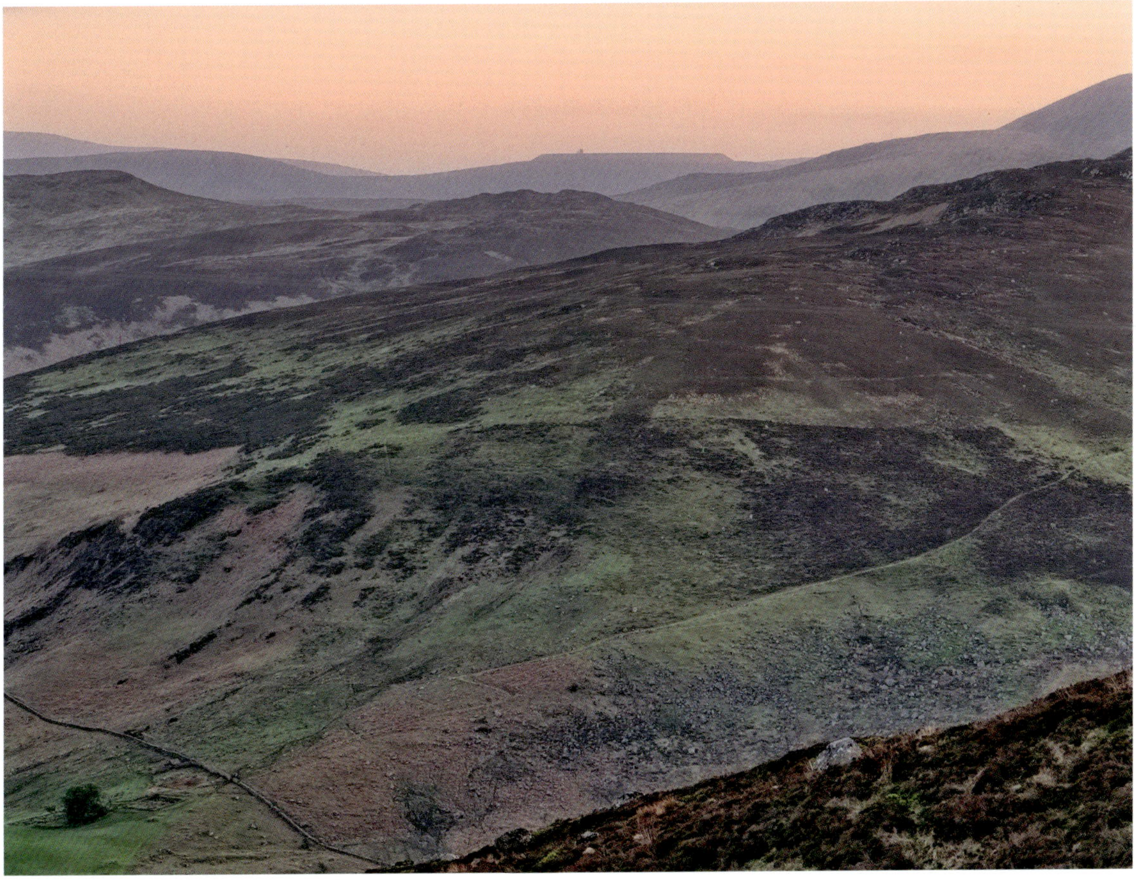

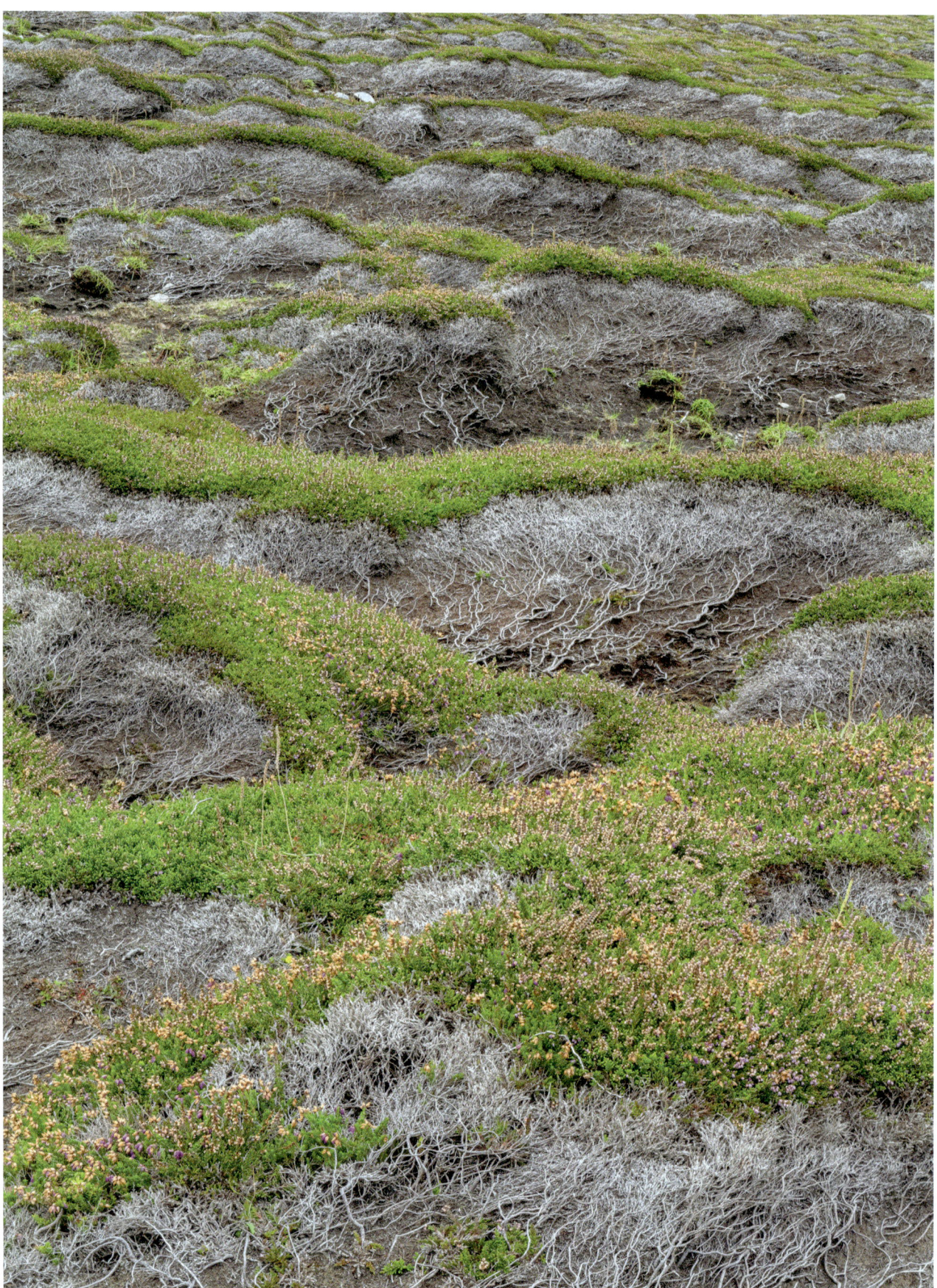

Autumnal montane heath

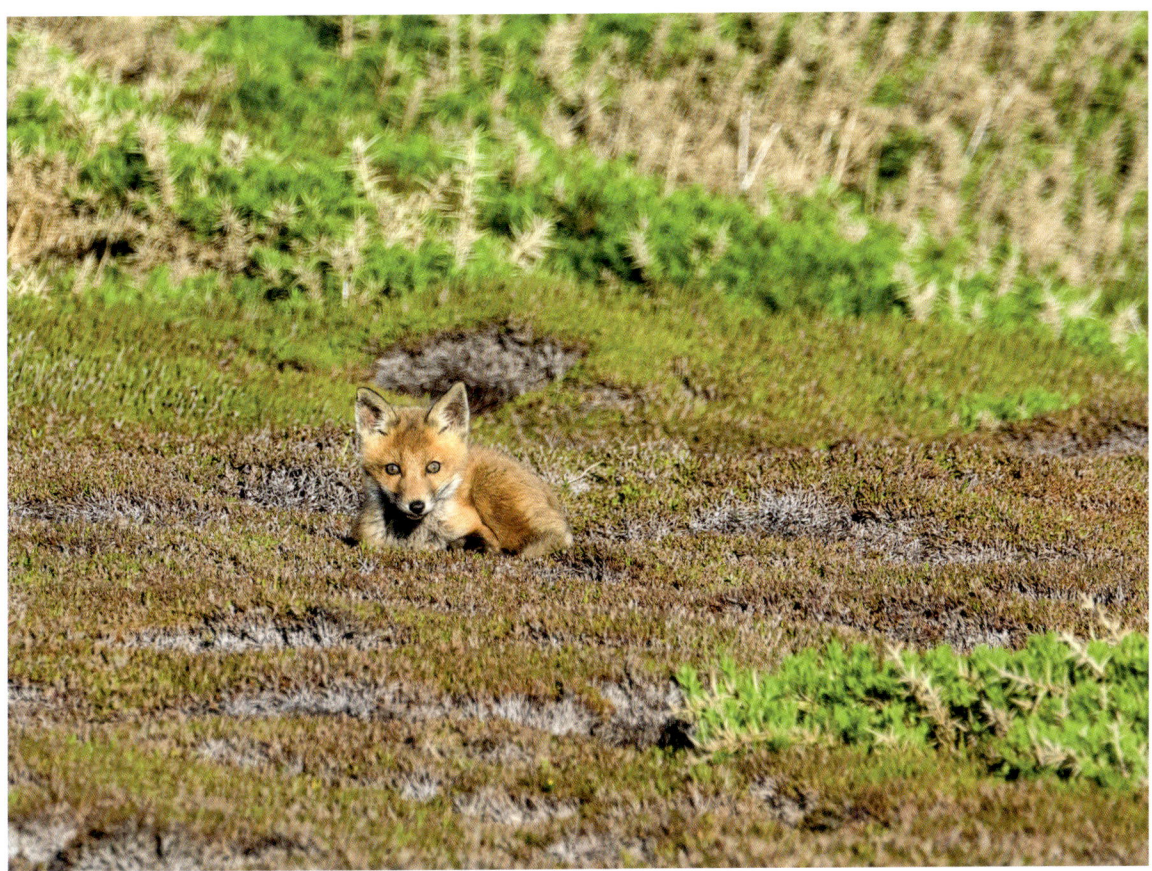

Fox cub resting near its den on montane heath

Dry silicious heath and wet heath commonly share the landscape with blanket bog and thrive in spots that are sufficiently drained and become waterlogged rarely or not at all. Dry silicious heath develops in the driest of those areas and features ling and bell heather, which are accompanied locally, in County Galway and County Mayo, by St. Dabeoc's heath. Western gorse, crowberry, bearberry, heath bedstraw and tormentil are also common, and in coastal regions spring squill, sheep's-bit, thrift and sea plantain can join the above as part of the dry silicious heath flora.

Wet heath is in appearance very close to blanket bog. It forms in places that are not permanently waterlogged and on steep upland slopes where peat formation is limited due to inclination. Depending on the underlying topography, however, wet heath can very well develop into blanket bog over time, which makes the two habitats even more difficult to identify. Indicator species for wet heath that are absent from dry silicious heath are purple moor grass, deergrass and cross-leaved heath. Species that occur in wet heath but not in blanket bog are heath rush and green-ribbed sedge. Black bog rush, on the other hand, indicates blanket bog and is missing from wet heath.

A variant of dry silicious heath and wet heath characterised by its location and rugged appearance is montane heath. Montane heath occurs in extreme environments, either at high altitudes or at the coast, where the soil or peat experiences constant and significant erosion due to rain and wind.

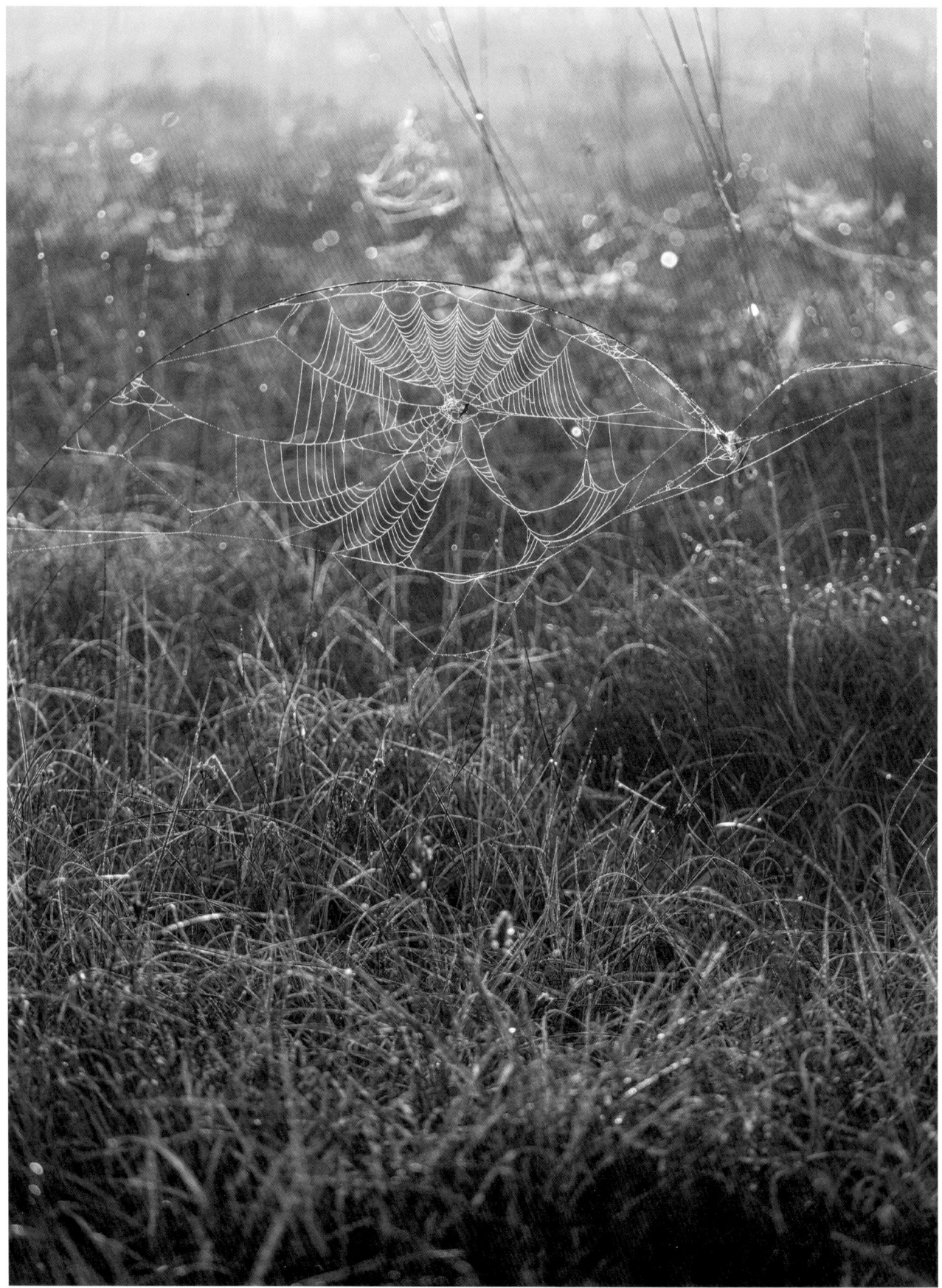

Spider's web at dawn

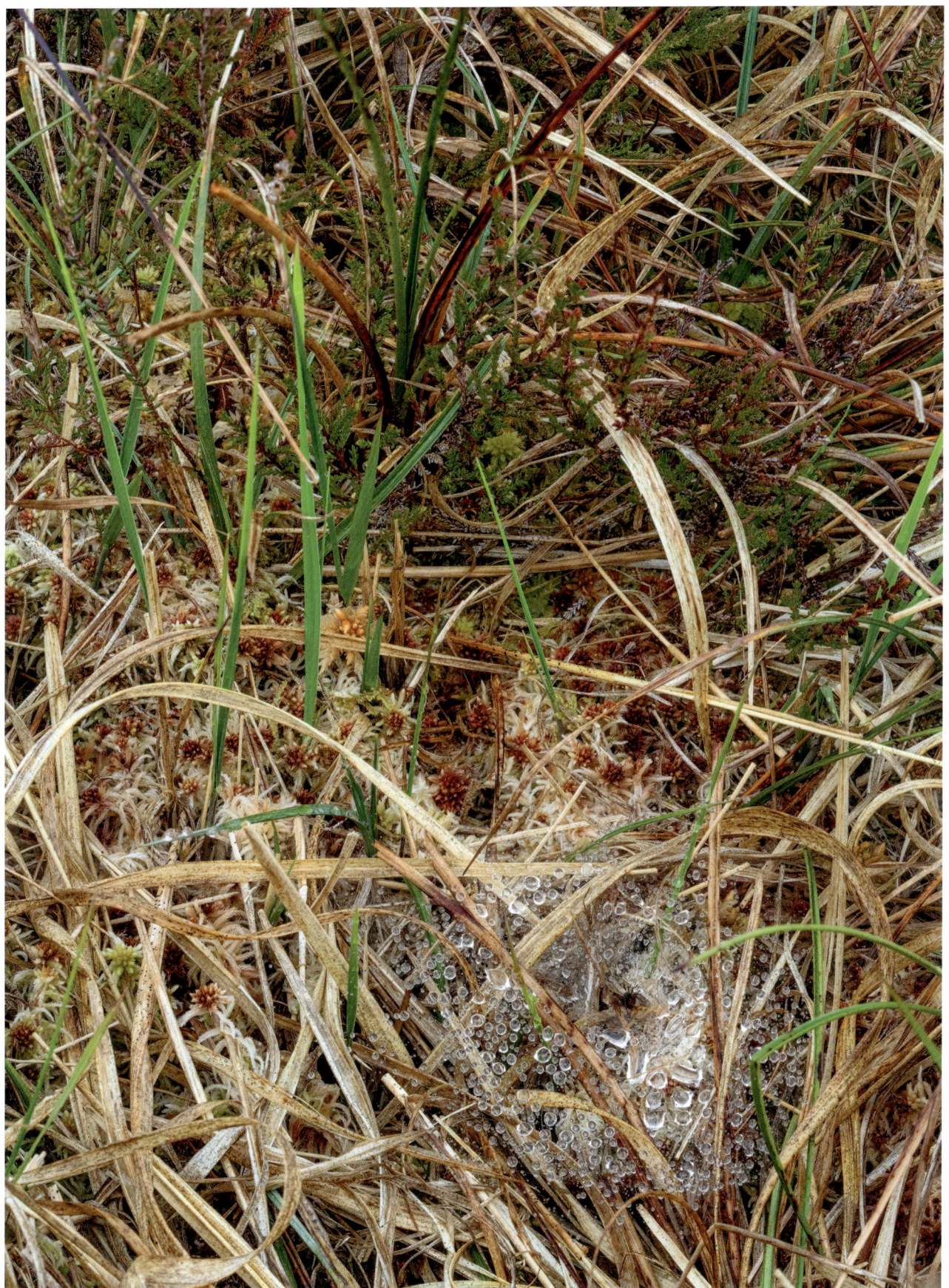

Funnel web spider

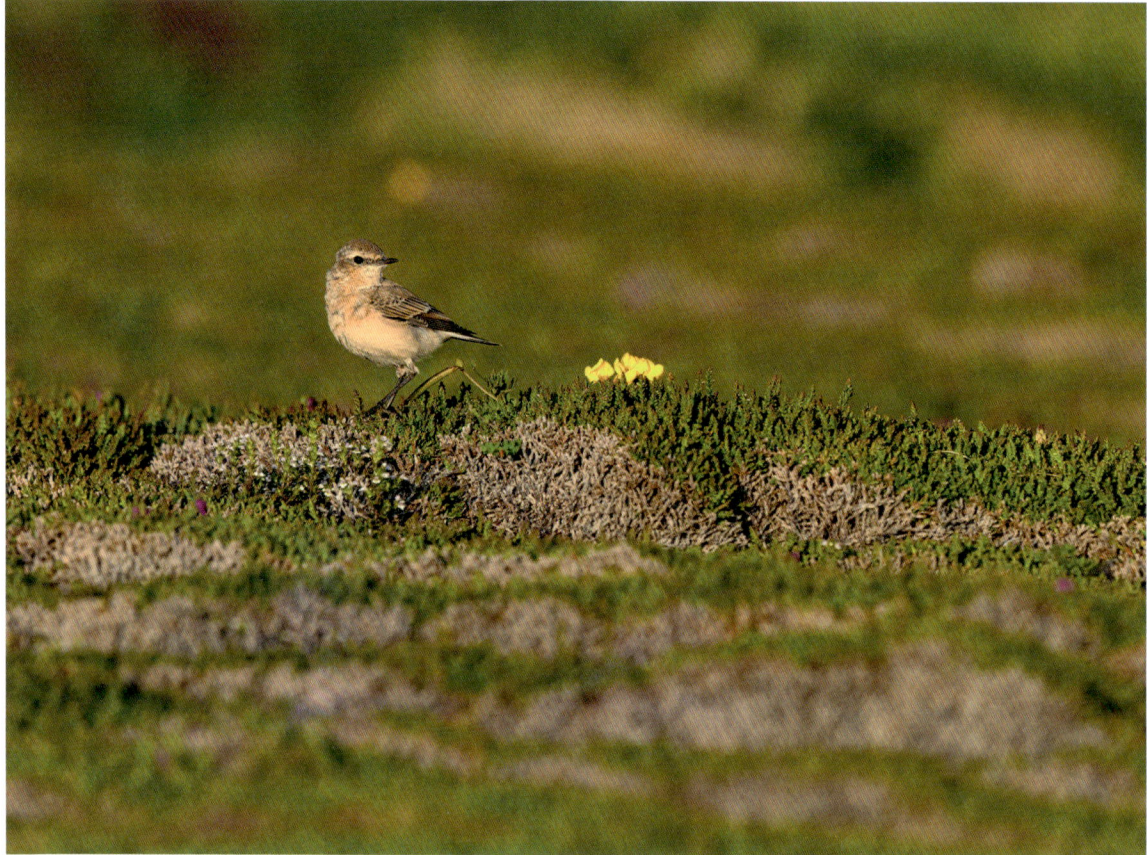

Wheatear

While Ireland's western peatlands have escaped the systematic and large-scale destruction the raised bogs had to endure, they were nevertheless decimated. Being harvested for fuel or transformed into farmland and commercial forestation changed most of the vast blanket bog areas – it is estimated that only 28% remain fully intact. One of those areas is the Roundstone Bog complex in Connemara, which stretches between Errisbeg Mountain and the peaks of the 12 Bens. This is a wide and open landscape covered in swathes of grasses, rushes and sedges and dotted with dark pools, ponds and lakes. Here and there, immense granite boulders and rocky outcrops emerge from the swaying grasses like islands.

A narrow road running straight east to west transects this forlorn scene. This road was once the main connection between Galway City and Clif-den, Connemara's main town at the most westerly edge of the area. It was the road any traveller visiting Connemara had to take. An old and gnarled hawthorn standing beside an elevated stretch of the road marks the spot of the infamous Halfway House, once a place to eat, rest and spend the night for the weary wayfarer.

How much of the tale surrounding the tavern is true is impossible to say, but as the story goes, in the late 18th century, the Halfway House was run by two sisters and three brothers who bolstered their income by letting some of their guests disappear and helping themselves to their belongings. Over the course of a few decades, numerous travellers vanished, and when not a hint of them was found, it was assumed that some kind of accident had befallen them on their journey through the treacherous bog. This was until one very hot

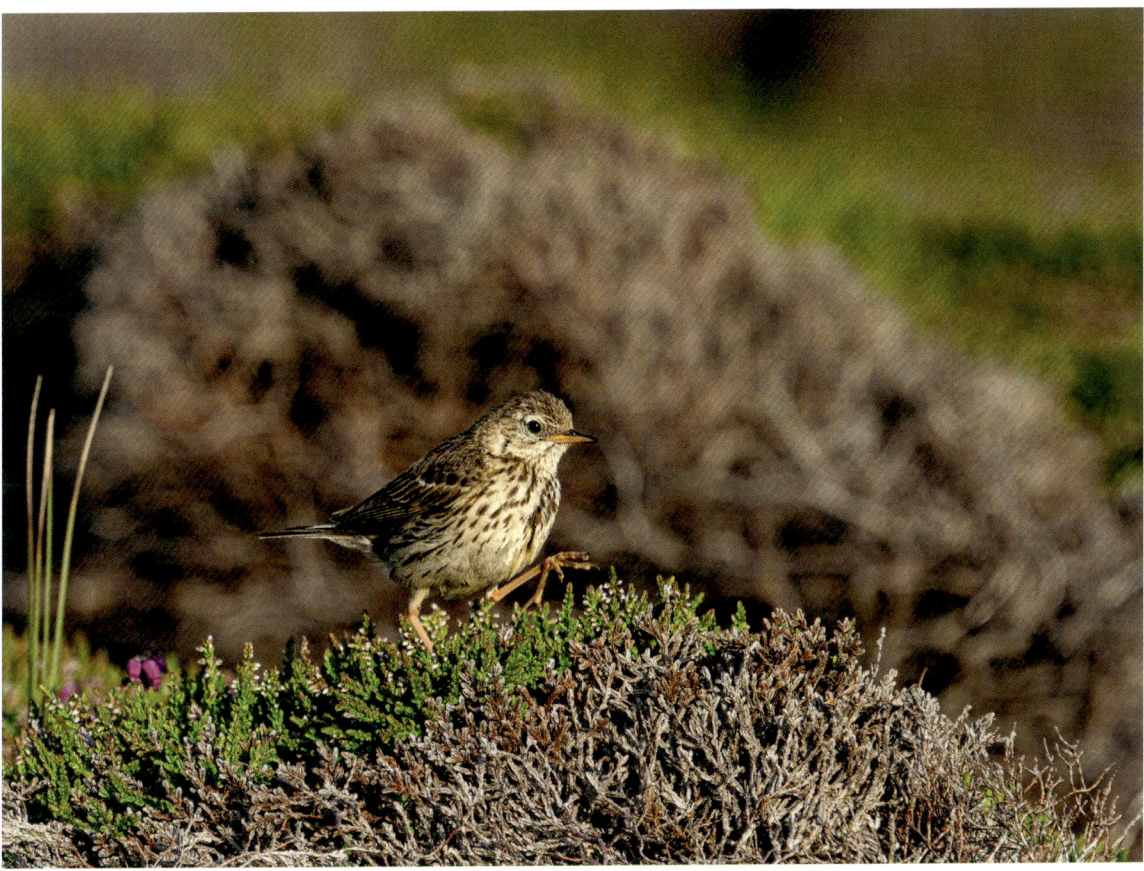

Meadow pipit

summer when a group of children made a gruesome discovery in one of the bog lakes not too far away from the tavern. Due to a long dry spell, the water levels of the lake were unusually low, uncovering countless human bones – some of them still partially covered by the remains of clothing. The investigation that followed quickly led to the owners of the Halfway House, who coincidently at the time were quarrelling among themselves about a particularly valuable item taken from their latest victim. It didn't take much effort by the constabulary to get to the truth, and the siblings soon sold each other out. The trial was swift and all five were hanged in Galway. How many victims the brothers and sisters had claimed over the years is unknown, but one legend says that each tree standing on the island in the lake where the victims were found represents one body.

The islands on the larger bog lakes at Roundstone Bog are indeed very different in appearance to the surrounding landscape. Many of them are densely covered with trees and shrubs, some with up to thirty-nine different species including rowan, birch, holly and yew. In addition to those woody plants, 132 herbs have been recorded on these islands, all anything but typical peatland plants and a stark contrast to the comparatively rather bleak blanket bog vegetation.

Still, in summertime the blanket bog flora delights with some wonderful shapes and colours. There are carpets of yellow bog asphodel, red sprinkles of sundews, vibrant green tussocks of purple moor grass and woolly tufts of common cotton grass as well as some rare oddities. St. Dabeoc's heath has its main distribution on the Iberian Peninsula and also makes sporadic appearances on

"

NATURALIST ROBERT LLOYD PRAEGER DESCRIBED THE AREA AS 'THE VERY LONELIEST PLACE IN THIS COUNTRY'

"

the northern coast of France and southern England. Its only other stronghold in northern Europe is Connemara, specifically the Roundstone area and the south of County Mayo, where it can be found in the drier areas of blanket bog and heath. How this plant ended up in the west of Ireland is still being debated, but it seems likely that it was introduced by pilgrims or other travellers. The same question of origin hangs over two other heather species. Mackay's heath, which produces no viable seed and instead spreads by layering, can only be found in Spain and a few spots on Ireland's west coast. Dorset heath, as the name suggests, is a common sight in Dorset and adjoining counties in England but also makes an unexpected appearance in only one known location at Roundstone Bog.

Journeying north from Roundstone towards Killary Harbour and Leenaun, onwards through the eerie Doolough Valley and around Clew Bay, the traveller will eventually reach a place that naturalist Robert Lloyd Praeger described as 'the very loneliest place in this country'. This area that stretches from the rugged coast, which is sheltered from the worst onslaughts of the Atlantic Ocean by the landmasses of Achill Island and the Doohoma Peninsula, to the ominous Nephin Beg range further inland indeed oozes loneliness and desolation. Within this area lies the Wild Nephin National Park, over 10,000 hectares that are supposed to become Ireland's first designated wilderness site. The long-term plan for

the area is to return the peatlands to their natural state, replace the existing Sitka spruce plantations with native woodland, and restore riparian habitats. How realistic this vision is, and how a planned system of walking trails fits into this wilderness, remains to be seen.

On the western side of the Nephin Beg mountains lies Owenduff Bog, which might just be Ireland's largest unspoiled expanse of blanket bog. Although named a bog, Owenduff is rather a tapestry of peatlands. There is lowland blanket bog, with an undulating cover of purple moor grass, black bog-rush, deergrass and cross-leaved heath and the rare marsh clubmoss, slender green-feather moss and bog orchid. Here and there, nutrient-rich flushes support the very rare marsh saxifrage and shining sickle moss. A few lakes and the Owenduff River also provide islands with a less acidic environment in the peatland landscape, which attracts species like common spike rush and bulbous rush, and in places where grassy floodplains have developed, plants like self-heal, bog pimpernel and the rare ivy-leafed bellflower thrive.

In the foothills of the mountains, upland blanket bog mixes with patches of wet heath, dry silicious heath and upland grassland. Around the summits, montane heath prevails, featuring ling, bell heather, crowberry and bilberry, and in rocky areas, the rare mountain species purple saxifrage and alpine meadow-rue survive.

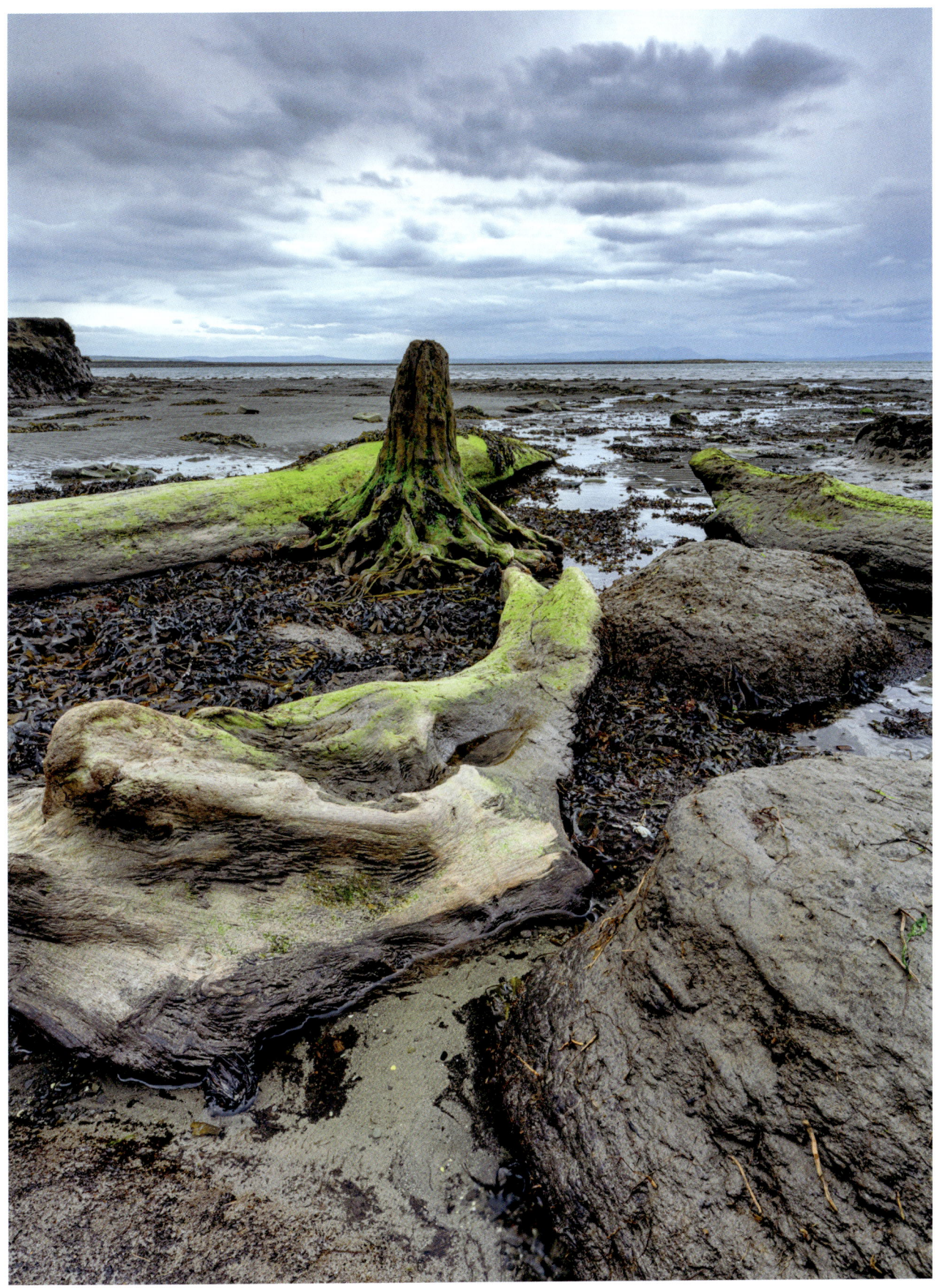

The Drowned Forest, County Clare

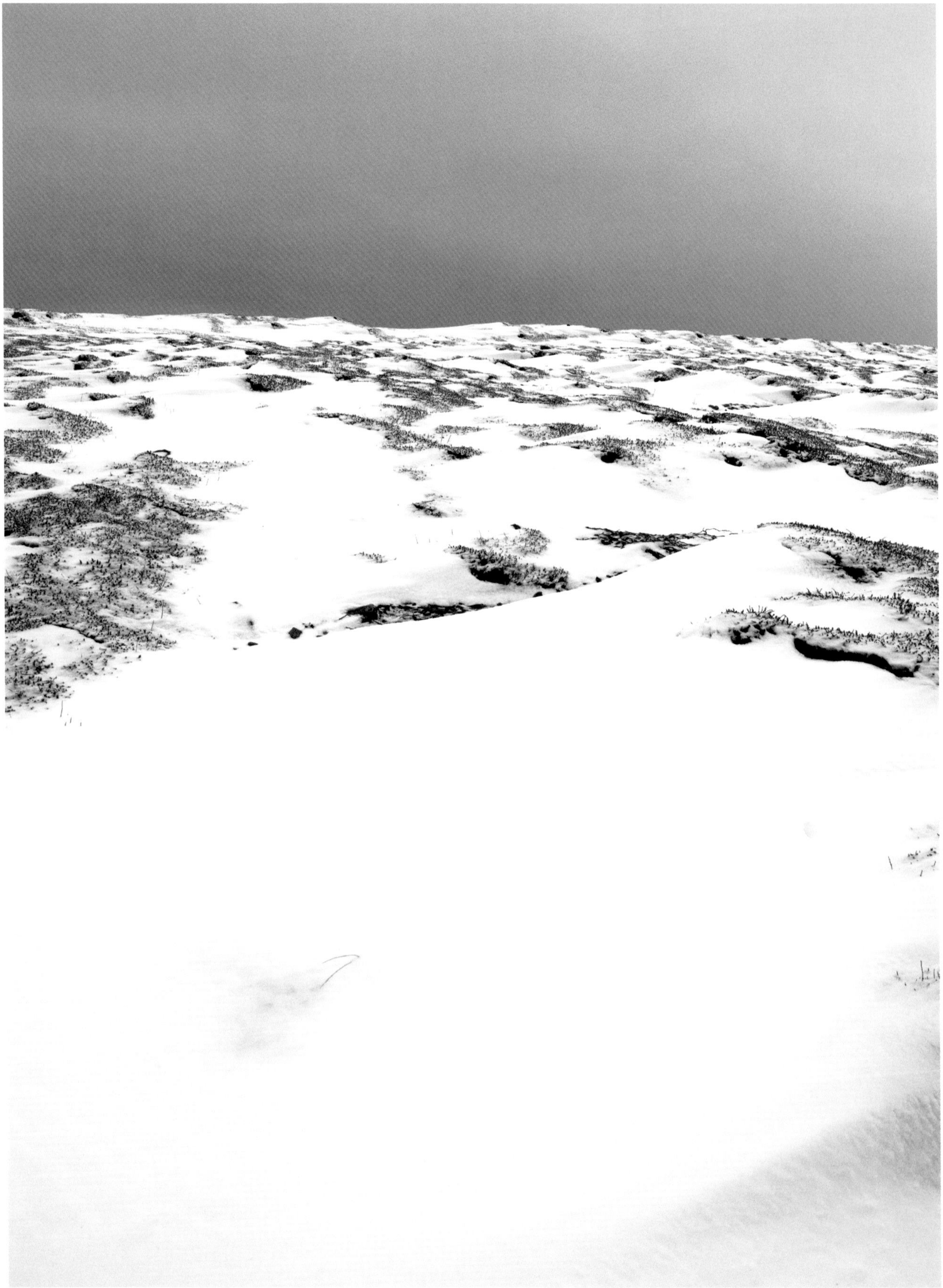

Montane heath in snow

"

EACH COMING AND GOING OF THE TIDE EXPOSES A MUCH OLDER LANDSCAPE, THE REMAINS OF ANCIENT FORESTS

"

The area is also one of the few breeding sites of the golden plover in Ireland and one of the last wintering places for the Greenland white-fronted goose on the west coast.

The peat in the wider Nephin area not only extends right up to the shoreline, in many places it stretches out into the Atlantic Ocean, providing the substrate for one of Ireland's most vibrant habitats, the saltmarsh. Saltmarshes are not classified as a peatland, but in many places along the west coast, saltmarsh flora has developed on underlying peat. In these places, the peat hosts plants it wouldn't normally support featuring characteristic coastal wildflowers like sea plantain, thrift, sea aster, scurvy grass and sea lavender alongside rushes and grasses like saltmarsh rush and red fescue and even sea-weeds. Turf fucoids, species of brown algae which thrive in miniature form on the peat, were first discovered at Bellacragher Bay in County Mayo, just south of Owenduff.

This coastal peat is a temporary landscape. Each coming and going of the tide scrapes away pieces of peat, shrinking this habitat and in the process exposing a much older landscape, the remains of ancient forests; tree trunks embedded in peat speak of a time when sea levels were lower and the climate less humid, less mild and less windy. Remains of these old forests can be seen all along the west coast. One particularly impressive example is located at a small bay in the Shannon Estuary in County Clare. Here, the peaty coastline is littered with pine and birch trunks and fallen logs, some still completely embedded in the peat, others partly exposed, and yet others freed from their peaty prison and deposited above the high tide line. These old forests can also be encountered further inland in other places where peat is slowly eroding. The shores of Lough Mask in County Mayo are prone to regular winter flooding, which has exposed wide areas of tree trunks protruding out of the peatlands surrounding the lake.

While blanket bog is a dominating habitat in the west of Ireland, it is rather scarce in the east. This is mainly due to significantly lower amounts of precipitation. Here, only mountainous areas like the Antrim Plateau, the Mourne Mountains and the Wicklow Mountains were able to capture enough moisture to grow and sustain blanket bogs.

"

ONE OF IRELAND'S RAREST BIRDS, THE RING OUZEL, IS ALSO KNOWN TO BREED IN THE UPPER REGIONS OF THE WICKLOW MOUNTAINS

"

The Wicklow Mountains, just south of Dublin, are the largest continuous upland area in Ireland, rising on average 300 to 600 metres above sea level and covering around 500 square kilometres, of which 205 square kilometres are today protected as the Wicklow Mountains National Park. While remnants of native oak forest still thrive in the valleys, and the highest peaks impress with vast moraines and rocky slopes, most of the area is covered in peat. Vast expanses of upland blanket bog and dry silicious and wet heath make the Wicklow Mountains the largest peatland area on the east coast. Due to climate conditions and human influences, peat depths are on average shallower than on the west coast and rarely go beyond two metres. The best preserved and actively growing peatland in the area is the Liffey Head Bog, which is a stronghold of the red grouse and some rare wildflowers like the small white orchid and bog orchid. One of Ireland's rarest birds, the ring ouzel, is also known to breed in the upper regions of the Wicklow Mountains, and although not a typical bird of bog and heath, the black bird with its characteristic white crescent on the chest uses the Wicklow Mountains peatlands to forage for food.

Similar to their western counterparts, most of the peatlands in the Wicklow Mountains are not in good shape. In addition to drainage, afforestation and peat harvesting, they have suffered from overgrazing from sheep and deer, which makes the peat more susceptible to erosion. In 2022, the National Parks and Wildlife Service joined forces with ReWild Wicklow volunteers on a trial project that aims to restore blanket bog flora: they erected timber dams, installed sheep fencing, and spread the site with heather mulch, seeds and fertiliser. This will in time bring these peatlands back to an actively growing state so that maybe in a not too distant future the words of Robert Lloyd Praeger will again ring true: 'You can set foot on the heather six miles from the centre of Dublin, and save for crossing two roads, not leave it till you drop down on Aughrim, thirty miles to the southward as the crow flies.'

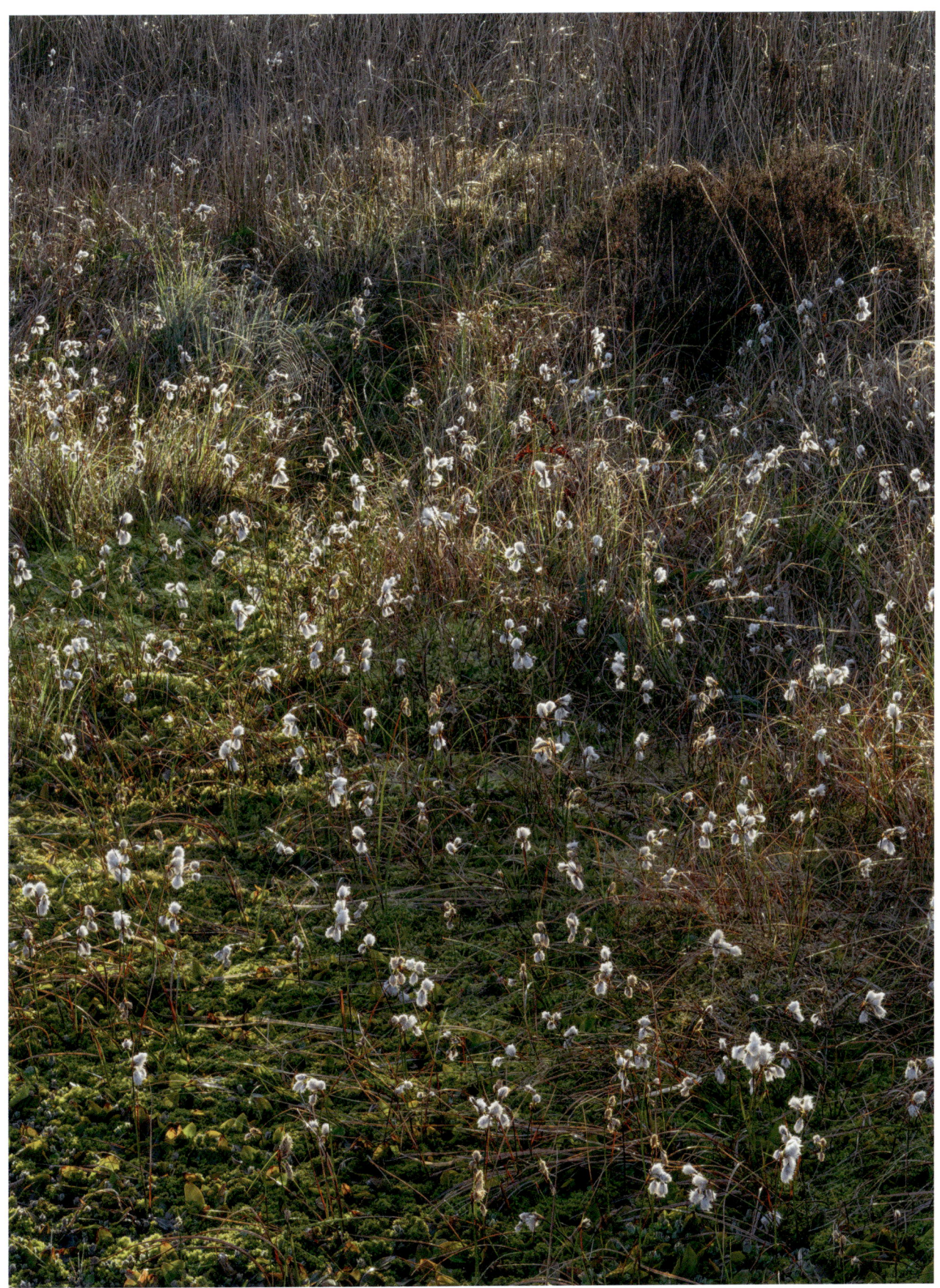

Hare's-tail cotton grass

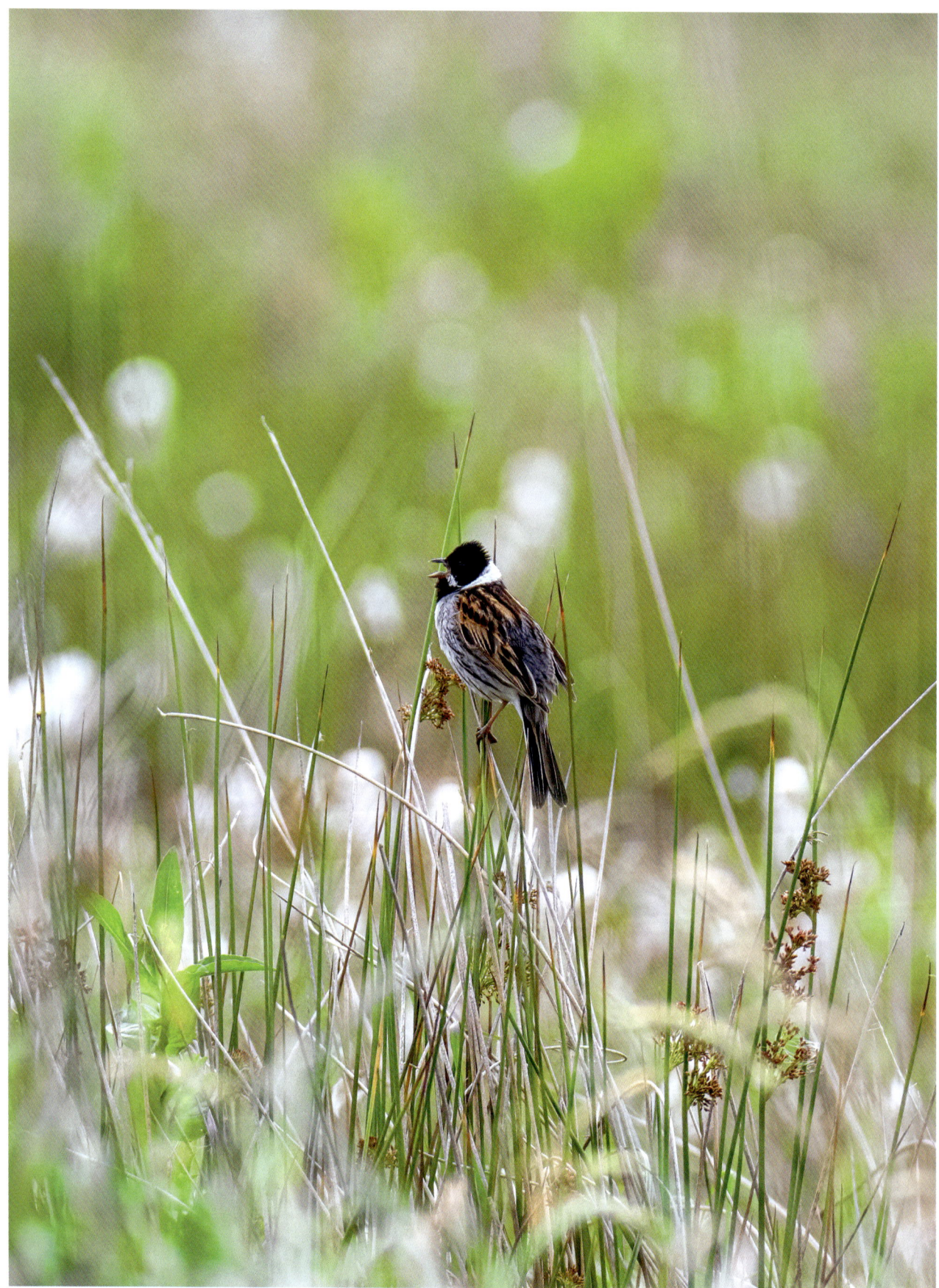

Reed bunting

Chapter Four
FLORA & FAUNA

Red and green carpets of sphagnum mosses, verdant varieties of grasses, purple clusters of heather, the yellow spikes of the bog asphodel and the red inflorescence of a marsh orchid, the tiny white flowers of the sundews and the purple ones of the butterworts, the crimson blossoms of the marsh cinquefoil and the fluffy seedheads of the cotton grasses. Without plants, there would be no peatlands. The lush, entangled vegetation we see on peatlands doesn't only grow there; the vegetation *is* the peatland.

The way plants have developed strategies and the ability to survive in the waterlogged, acidic and nutrient-poor environment of the bogs and heaths is something to be marvelled at. The vibrant communities of mosses, lichen, grasses, rushes, sedges, shrubs and other flowering plants have built a multitextured and colourful environment like no other. They have found intriguingly clever ways to not only thrive in their challenging surroundings but also provide a home for a small but alluring group of mostly temporary resident invertebrates, amphibians, reptiles and birds. Above the luscious carpet of peatland flora hover dragonflies and damselflies, butterflies

dance in the light breeze, and in the shadows at the edge of a pool sits the raft spider, motionless, waiting for its prey. The skylark sings high up in the sky, the drumming of the snipe echoes in the distance, and a rustling in the heather could have been caused by the elusive red grouse. In spring, frog spawn appears in the ponds, and in summer the viviparous lizard can be seen sunbathing on rocks.

While many peatland plants thrive exclusively in this very particular habitat, most of the animals that are encountered here are not restricted to peatlands and can also be found elsewhere. The reasons they come, however, are the plants. The peatland flora provides shelter and food, and some animals, like the red grouse, have indeed formed a strong bond with certain plants. It is not only the plants themselves that matter, but also how they grow. If the plants are too far apart, some species are not able to move through the peatland; if they grow too dense, other species don't find enough room to build their nests. A peatland might look chaotic to us, but for the plants and animals living in it, everything is well arranged and in a certain order – there is method to the madness.

"

IRELAND'S PEATLANDS ARE HOME TO FOUR GROUPS OF CARNIVOROUS PLANTS: SUNDEWS, BUTTERWORTS, BLADDERWORTS, AND PITCHER PLANTS

"

FLORA

To live comfortably, most peatland plants have adopted a very slow growth rate to reduce their need for nutrients. Some carry evergreen leaves to avoid the need to produce new ones every year, which also greatly reduces the required nutrient intake. Some have developed underground storage facilities in their roots and rhizomes they can fall back on in times of need, and others have the ability to absorb nutrients and carbon dioxide directly from the surrounding water into their leaves. Some, including most rushes and sedges, grow very long roots to reach hidden nutrients deep inside or even below the peat layers, while others, like the heathers, stretch out their roots just beneath the surface to catch any sustenance as soon as it lands on the ground. Some plants became even more sophisticated and entered partnerships with fungi or bacteria that provide nitrogen in exchange for carbohydrates and oxygen, while others just tap into the roots of their neighbours to fulfil their nutritious needs. The most innovative way to gather nutrients, however, has been fashioned by the carnivorous plants.

Carnivorous plants

This family of plants has successfully repurposed a defence mechanism widely used in the plant world and made it into a digestion tool that allows them to extract nutrients from insects. Chitinase is a chemical compound many plants produce to fight fungal infections. This compound breaks down the cell membrane of fungi, which is made of chitin, effectively killing the invading fungus. Coincidently, chitin also makes up the exoskeleton of insects, and the plants who figured this out made a vast new food source available to them.

Ireland's peatlands are home to four groups of carnivorous plants: the sundews, the butterworts, the bladderworts, and the pitcher plants.

Pitcher plants are not native to Ireland but nevertheless seem to feel rather comfortable in our peatlands. The Canadian pitcher plant was introduced into a bog in County Roscommon in 1906 for reasons that can only be speculated upon. Since then, the plant has colonised several other locations in the country. Whether this was by itself or with some help is not known for sure.

Unlike Ireland's native carnivorous plants, pitcher plants are rather easy to spot. They have repurposed their leaves into tubular pitchers or funnels that reach some fifteen centimetres above the bog surface and, with their characteristic red and purple vein markings, stand out from their surroundings. The individual pitchers hold a mixture of rainwater and digestive juices and lure insects and other small invertebrates with their vibrant colours and a promising scent. The excited

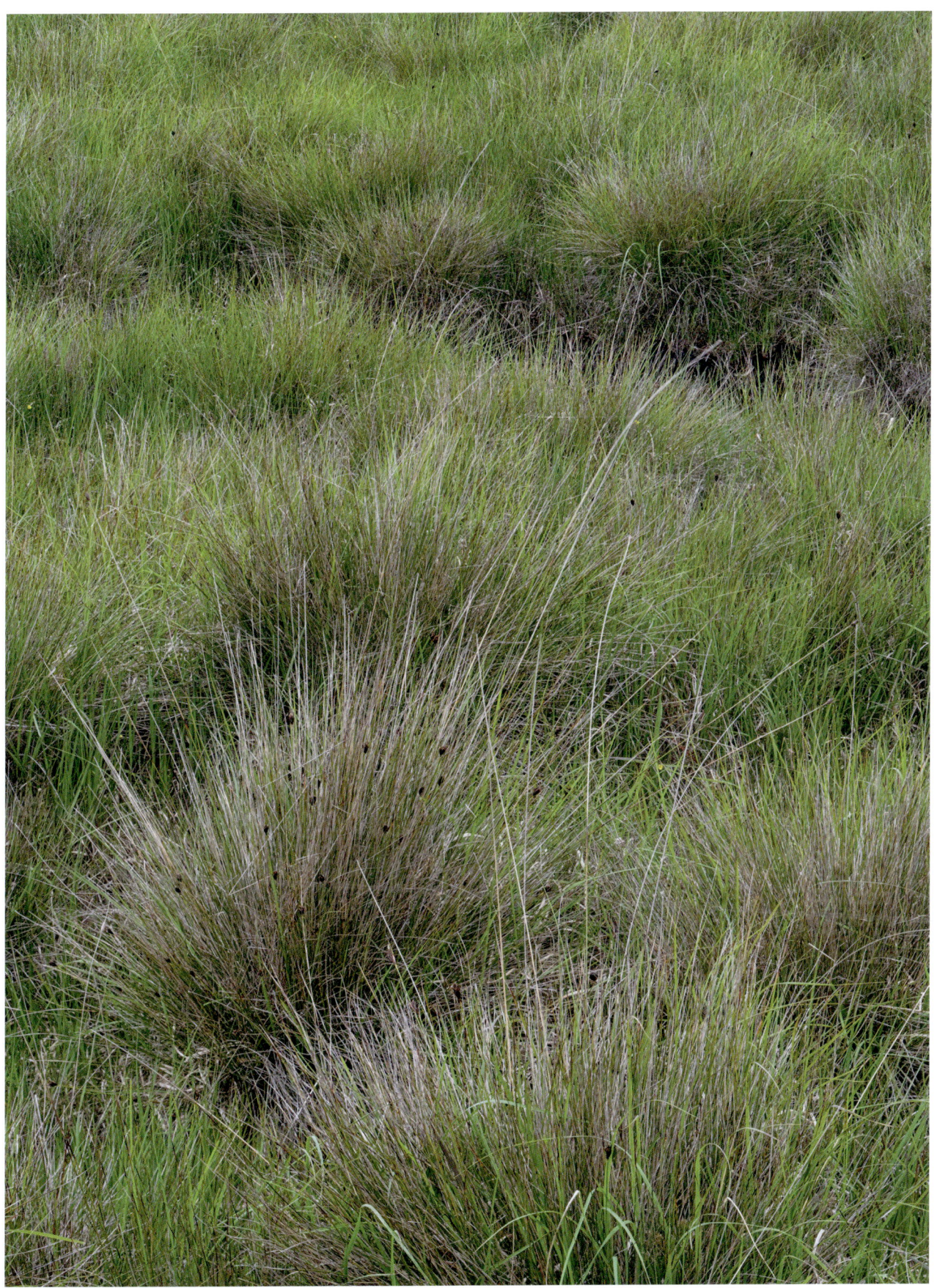

Purple moor grass

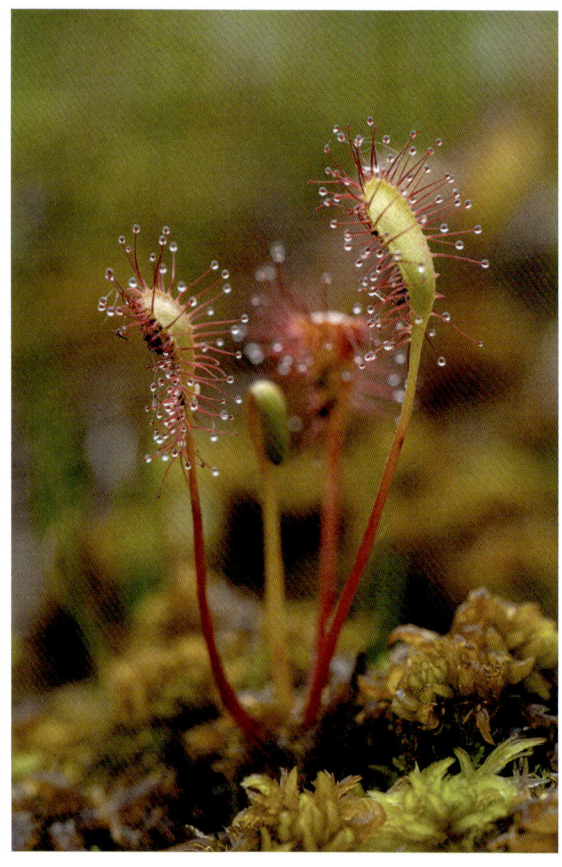

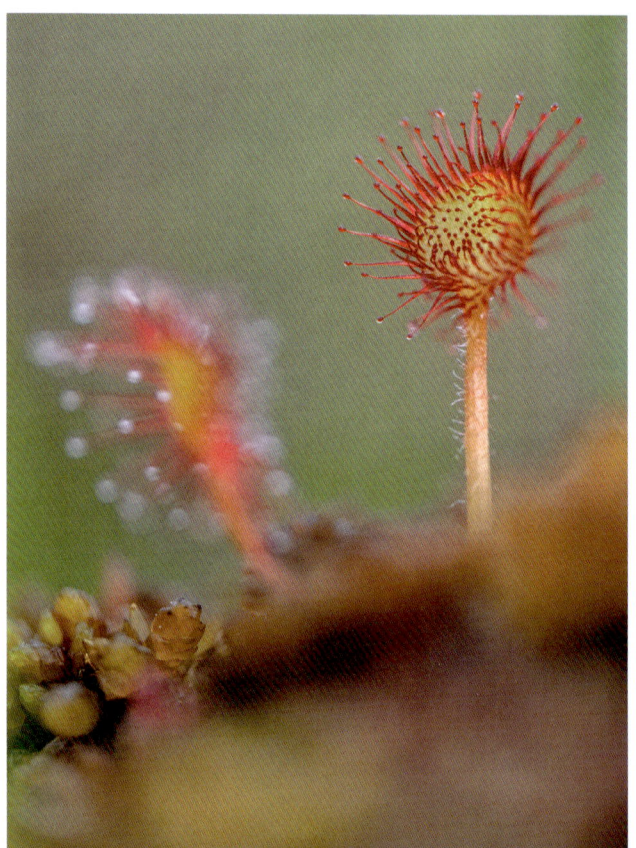

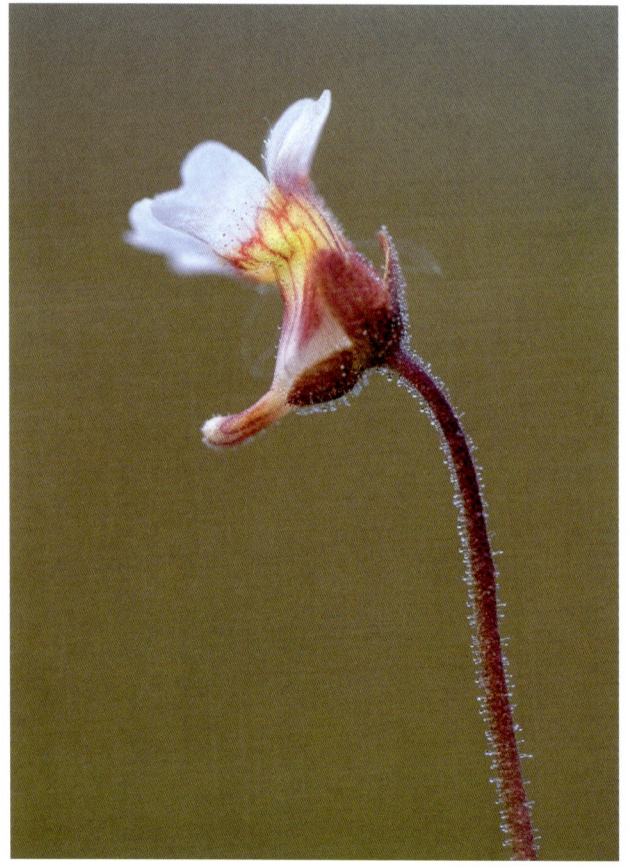

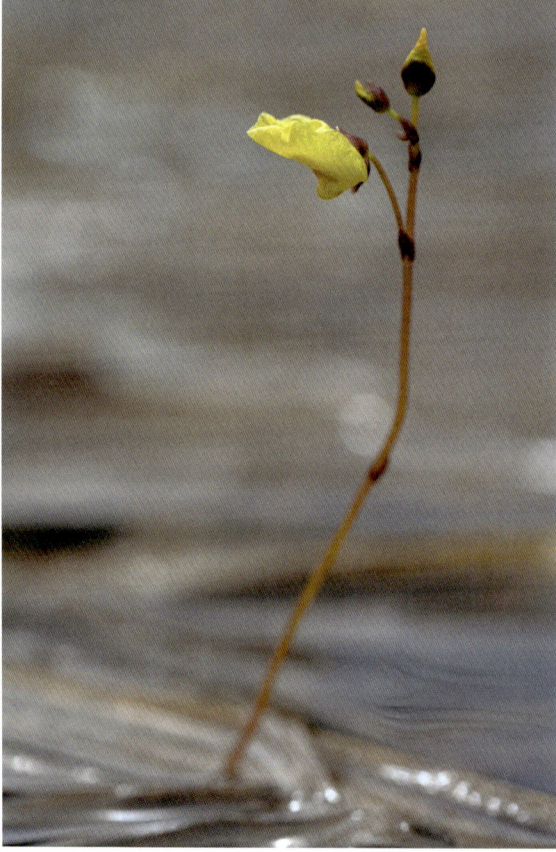

"

ALL THREE SUNDEW SPECIES HAVE REDESIGNED THEIR LEAVES AND TURNED THEM INTO STICKY INSECT TRAPS

"

animals land on the top of the pitcher, but instead of finding a tasty meal, they lose their footing on the slippery surface of the leaf and tumble into the depths of the pitcher from which there is no escape. Pitcher plants catch a wide range of invertebrates. In just one single pitcher, a study discovered 205 prey items; mainly mites but also flies, midges, beetles, small wasps and even spiders.

Less easy to spot are tiny sundews, which can form tightly packed colonies in places. The three sundew species can be easily differentiated by their leaves. The great sundew (with oblong leaves up to four centimetres long) and the oblong-leaved sundew (with oblong leaves only up to one centimetre long) can be mainly found in the western half of the country, while the round-leaved sundew (the smallest of the three, with round leaves measuring less than one centimetre in diameter) is common in all peatlands.

All three sundew species have redesigned their leaves and turned them into sticky insect traps. The surface of the fleshy leaves features numerous red-tinged, stalk-like glands, each with a syrupy droplet – the 'dew' – on its top. Once an insect lands on the stalk, hoping for a tasty sip, a cascade of actions happens: The gland senses the movement of the insect and sends out a chemical message. In response to this message, the plant releases a hormone that weakens the cell walls of the leaf, while at the same time a protein redistributes the water

content of the cells, removing water from some and pumping it into others. These actions cause the leaf to bend inwards, and within minutes the insect is trapped for good. Over the following hours, the leaf continues to roll up, completely engulfing its prey, and chitinase and other digestive enzymes are released. In this manner, each sundew leaf can catch around five insects every month.

The sundews once played an important role in herbal medicine and have been used, chopped and squeezed into a viscose juice, to treat lung ailments like coughs and asthma. Today, the chemicals responsible for the anti-inflammatory and anti-spasmodic properties of the sundews are used in a variety of cough and cold medicines.

Just like the sundews, the butterworts got their name from the special adaptations of their leaves, whose glossy surfaces look like they have been smeared with butter. This characteristic appearance is even reflected in the Latin name of their genus: *Pinguicula* roughly translates into 'little greasy one'.

The aptly named large-flowered butterwort, a member of the Lusitanian flora, can be mainly found in the southwest; the similar looking but overall smaller common butterwort is at home mainly in the west and north. While these two species produce eye-catching purple flowers, the small white flowers of the pale butterwort, found all over Ireland, seem less notable. On closer inspection, however, the pale butterwort reveals a rather

Opposite top left: Oblong-leaved sundew
Opposite top right: Round leaved sundew

Opposite bottom left: Pale butterwort
Opposite bottom right: Lesser bladderwort

"

THE LEAVES OF THE BUTTERWORT ARE COVERED IN ONE OF THE STICKIEST SUBSTANCES KNOWN IN THE PLANT WORLD

"

intriguing beauty. The white petals show an ever so slightly pink or purplish hue, and the inside of the flower presents itself in a vibrant yellow.

The butterworts' hunting technique is not unlike that of the sundews, but while the latter have developed rather elaborate leaf traps, the butterworts keep things simple. Their elongated, fleshy leaves, which are spread out in a rosette on the ground, are slightly rolled up along the edges to prevent any prey from escaping but are not able to perform movements like the leaves of the sundew. They don't need to, because the leaves are covered in one of the stickiest substances known in the plant world. Any insect that makes its way onto the leaf is very unlikely to get away. It is not uncommon to see butterwort leaves littered with countless carcasses of tiny flies and other minute critters who got digested in the very spot they stepped on. Research into butterworts has shown that the sticky leaves not only capture insects but also pollen carried by the wind. Pollen is highly nutritious, and it is thought that the butterworts take advantage of that and are able to extract some extra sustenance from it.

Bladderworts are the aquatic representatives of the carnivorous plants, and Ireland is home to three species: the lesser bladderwort, intermediate bladderwort, and greater bladderwort. They spend their existence floating freely in pools and lakes and catch their prey in an unusual way, which also

gives them their name. Bladderworts feature small bladders sitting on short stalks off the main stem. These bladders are capped by a lid and equipped with sensitive trigger hairs on the outside. Whenever a small animal makes the mistake of getting too close to the trigger hairs, the lid opens, the vacuum inside the bladder sucks the unfortunate individual inside, and the lid snaps shut. All of this happens as quick as 1/15,000th of a second. After the victim has been digested, its remains are expelled and specialised glands suck out the water from the bladder to re-establish the vacuum. The bladder is now ready for the next meal.

For most of the year, the partly or fully submerged bladderworts are inconspicuous; only in June and July do they extend long stalks above the surface of the water and produce delicate yellow flowers that add some brightness and vibrancy to the dark bog pools.

Besides the bladderworts, a few other species manage to survive in the inhospitable pools and ponds of the bog and even do so in the traditional, not carnivorous, plant fashion. One of them is pipewort, a rare species confined to a few shallow pools in counties Galway, Donegal and Kerry, where it can be spotted by the keen eye when its round, tiny, greyish-white flowers reach above the water surface from July to September. While rare, the pipewort seems to be a very consistent plant. Pollen studies conducted at Roundstone

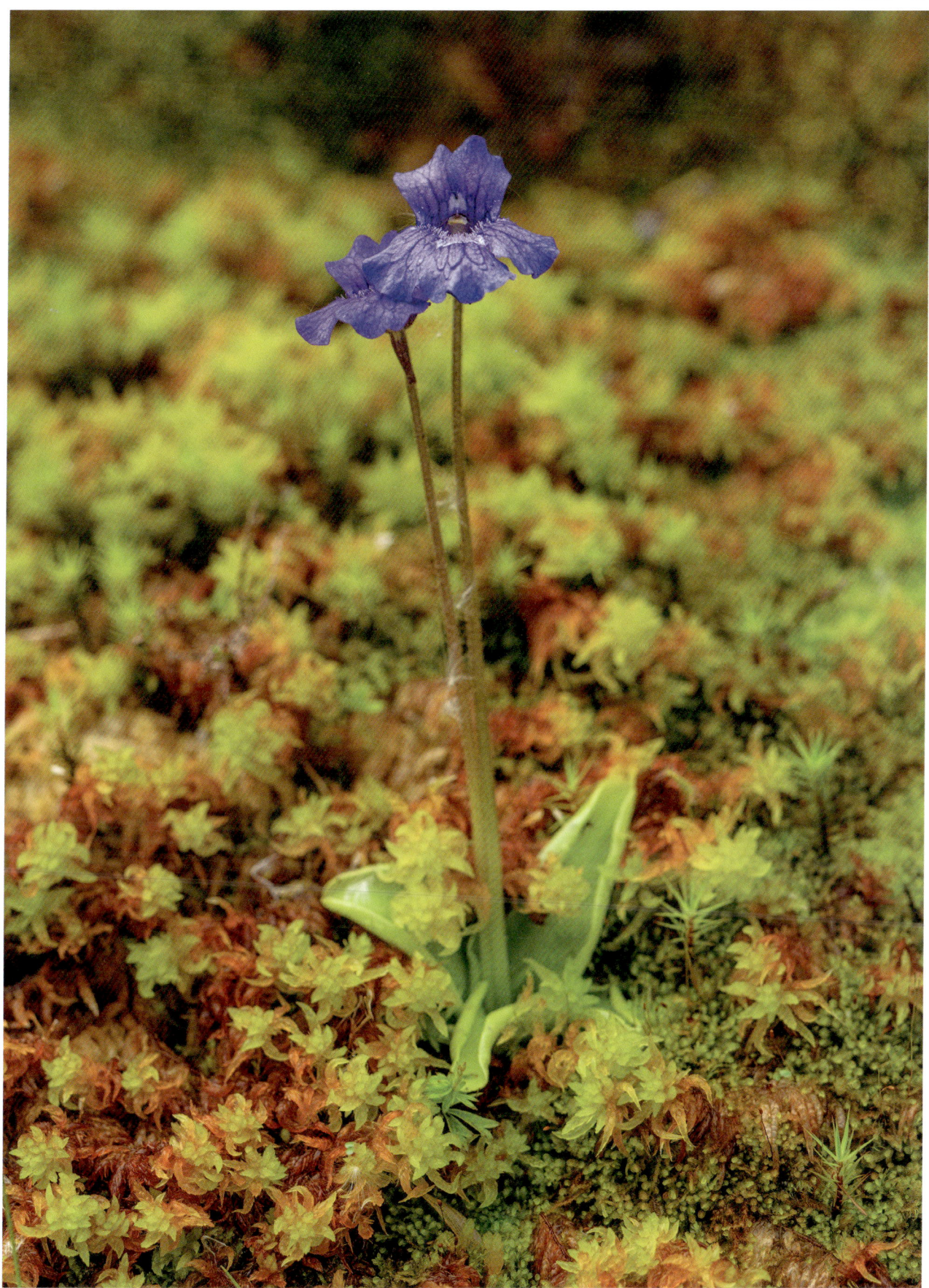

Large-flowered butterwort

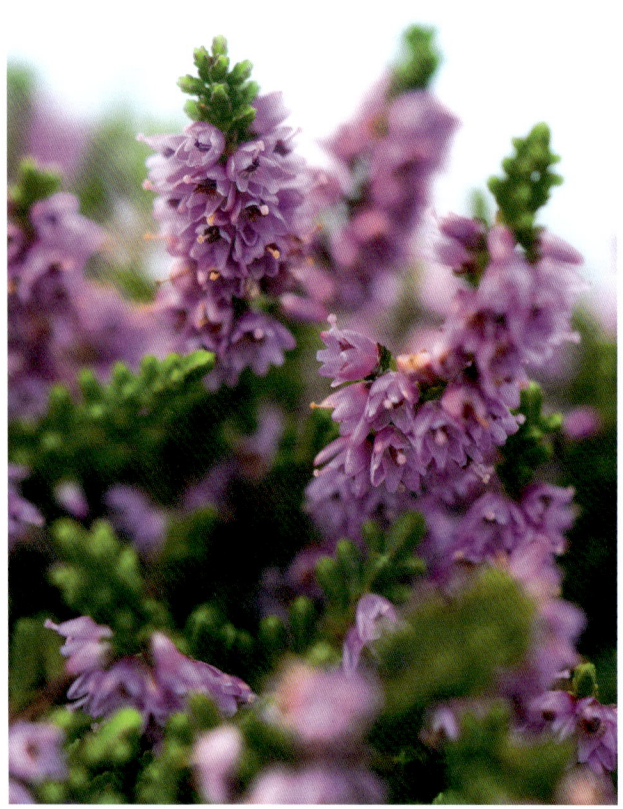

Ling (common heather)

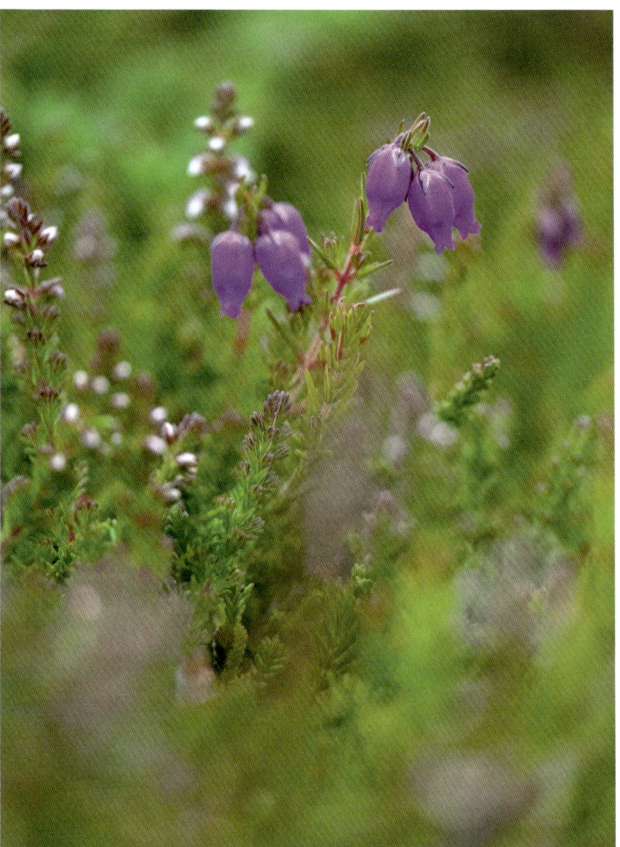

Cross-leaved heath

Bog have shown that pipewort has been present in the pools and ponds of Connemara for some 6,000 years.

The water lobelia, abundant along the west coast, also shows its flower in late summer but does so in a more prominent way than the pipewort. Its unmistakably elegant lilac-white flowers sit on long, unbranched stems, swaying up to twenty-five centimetres above the water surface.

White and yellow water lilies are not typical of peatlands and prefer a more hospitable environment, but they can be found in ponds and lakes that have a connection to groundwater or are fed by a stream or river.

Heathers and other shrubs

Heathers are among the most characteristic, widespread and, when they are in bloom in late summer, eye-catching peatland plants. These small evergreen shrubs are the main colonists on heaths and can form thick carpets and prominent hummocks on all but the wettest areas of raised and blanket bogs. Because of their abundance, all heather species, with their tiny but nectar-rich purple flowers, are a vital food plant for insects in the latter part of the summer and early autumn.

Not too long ago, it wasn't just the insects who had a huge appreciation for this plant. Heather flowers have been used to flavour beer – a practice likely introduced by the Vikings – make a fragrant tea, and produce honey. The woody parts have been made into brooms, ropes and baskets. Whole plants were once very popular as bedding, not only because of their softness but also for their aromatic smell, and in areas where reeds were not available heathers were used as thatch.

To gain their required nutrients, heathers live in partnership with mycorrhizal fungi. The fungal

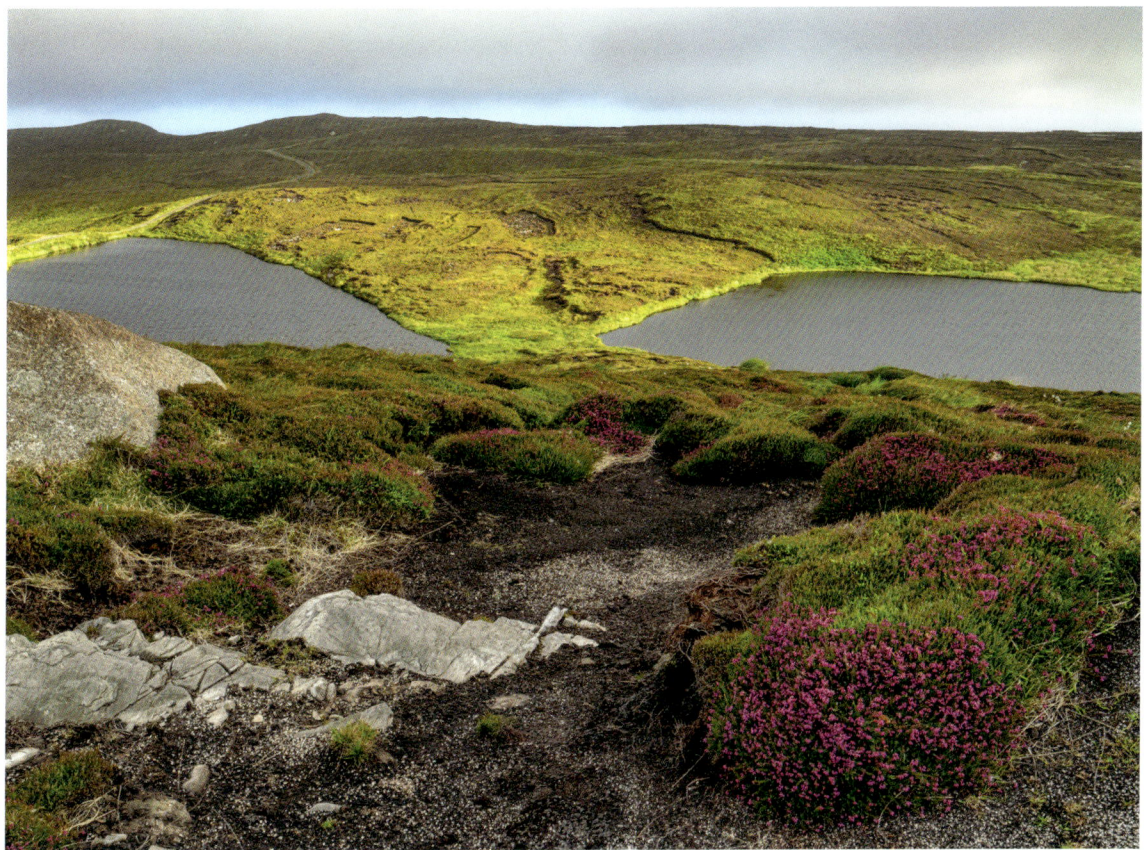

Blanket bog on Arranmore Island, County Donegal

partner lives coiled around the roots of its host and makes nitrogen and phosphorus available to the heather plant. The heather in return sends oxygen and carbohydrates to its fungal partner, and it is likely down to this long-established partnership that heathers were able to be so successful in conquering the challenging environment of the peatlands.

The most common heather species in Ireland are ling (also known as common heather), bell heather and cross-leaved heath. The latter two are very similar in appearance, but while cross-leaved heath prominently displays its bell-shaped flowers only on one side at the top of the stem, the flowers of the bell heather pop up all around the stem. Bell heather prefers the drier areas of peatlands, while cross-leaved heath has no problem with periodically waterlogged ground and therefore thrives mostly in wetter areas.

While these three heathers can be found all over Ireland, the west of the country hosts three more unusual and rare species: Irish heath (also known as Mediterranean heath), St. Dabeoc's heath and Mackay's heath all belong to the Lusitanian flora and are restricted to a few sites in Connemara and County Mayo. How these plants ended up in Ireland is still widely debated, but they seem to have been thriving here for a long time.

St. Dabeoc's heath, or *fraoch na haon choise*, could have been one of Ireland's early contraceptives. The Welsh naturalist Edward Lhuyd described in 1700 that women in counties Galway and Mayo carried sprigs of the plant to prevent, as Edward phrased it, 'mishaps' . The first part of its scientific name, *Daboecia cantabrica*, refers to St. Dabeoc, an elusive early Irish monk who might have been preaching in the 5th or 6th century and who is closely associated with Lough Derg in County Donegal, a medium-sized lake surrounded by peatlands.

"

THE LAST SUNDAY IN JULY WAS KNOWN AS FRAUGHAN OR HEATHER-BERRY SUNDAY

"

The monk allegedly used this heather species as a medicinal herb to cure infections, particularly those of the kidneys and liver, and to treat coughs and colds. The second part of the name refers to the Cantabrian Mountains which sit near the coast of northern Spain, where the plant grows on acid soils on cliffs and rocky shores.

Mackay's heath has a similar distribution on the European mainland to that of St. Dabeoc's heath, but it prefers a wetter environment. A study conducted by Galway University found that Mackay's heath, which so far has been found in only four locations in Ireland, always grows within one kilometre of tracks and roads near sandy bays. It is known that heather was used by early tradespeople as a packing material to protect delicate goods; this and the very localised and limited distribution of Mackay's heath supports the theory that the species might have been introduced accidentally by visiting traders from northern Spain. A pollen study conducted at Cleggan Mountain in Connemara placed the arrival of Irish heath as recently as the 15th century, so it is not unlikely that other Lusitanian species, including the three heathers, found their way to Ireland in the very same way. Another theory suggests that the plants might have survived the glaciation periods, which, due to their Mediterranean origin, seems unlikely. Or perhaps they travelled to Ireland via the short-lived land bridge after the end of the Ice Age, which is possible but raises the question of why the Lusitanian heathers are not more widespread.

The bog myrtle is a typical west of Ireland shrub that feels right at home in its acidic and nutrient-poor environment. It is the largest among all the shrubby peatland plants and grows up to one metre in height, locally forming small colonies that overshadow the undergrowth of heathers, grasses, rushes and sedges. The bog myrtle carries evergreen lanceolate leaves, which turn from a light to a dark green and eventually to a warm brown during their life cycle. The shrub is known for its distinctive sweet fragrance that emanates from the leaves and small yellow dots on the reddish branches. The origin of this scent is an oily resin, which is used as an essential oil and an ingredient in skincare products like soaps and lotions to this day.

While the resin of the bog myrtle is still rather popular, other uses of the plant have been widely forgotten. Once, its leaves flavoured soups, stews and even beer, while the dried fruits were a popular spice for breads and cakes. The bog myrtle was also known for its medicinal applications. Documents show that all parts of the shrub were used to treat sore throats, measles and kidney ailments, and today we know that bog myrtle contains high

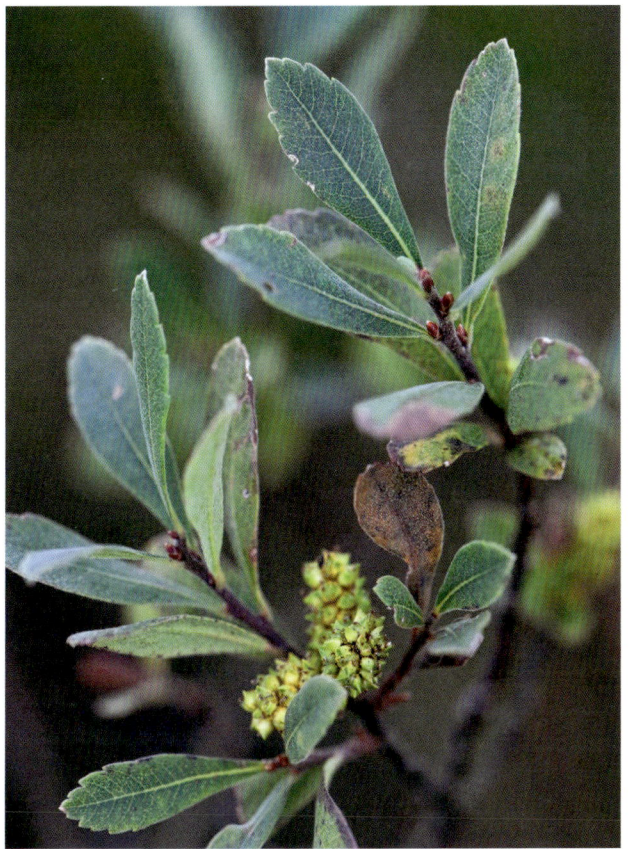

Bog myrtle

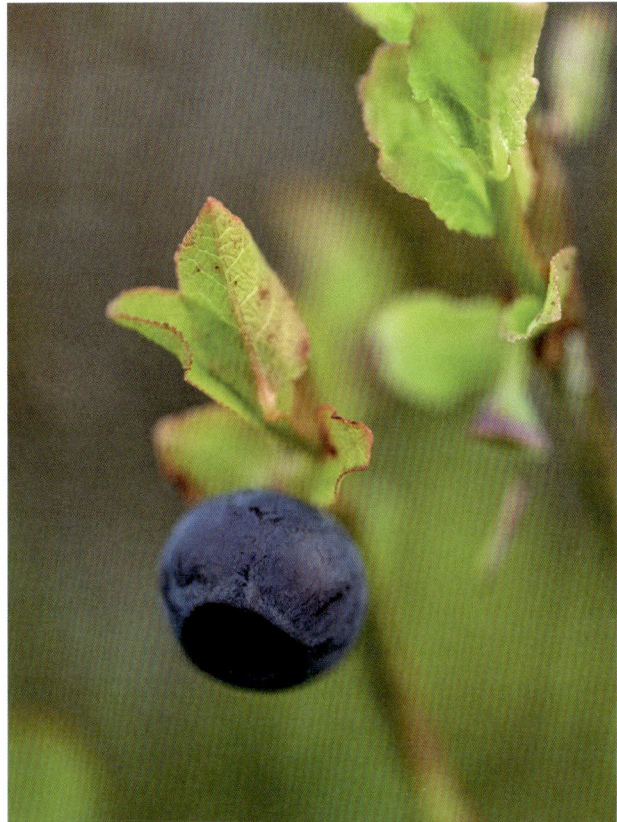

Bilberry

concentrations of tannins and flavonoids which have anti-inflammatory and antimicrobial properties. Another very welcome feature of the plant, especially when visiting the bog on a calm summer day, is its effectiveness as an insect repellent. The aroma of its crushed leaves keeps midges away very effectively, a fact that I can personally corroborate.

Also very popular in days gone by was the bilberry, the cousin of the domesticated blueberry. The bilberry grows as a small shrub on mountainside heathlands and in a slightly bigger version on raised bogs, and its former importance as a food staple is reflected in the variety of names by which it was known: whortleberry, blaeberry, huckleberry, whinberry, and in Irish, *fraochán*. Excavations in Dublin have exposed bilberry seeds dating back around a thousand years. As there were no bilberries growing in Dublin city centre, the fruits must

have been imported from the nearby Wicklow Mountains, which led to the theory that there once was a thriving bilberry trade on Ireland's east coast and perhaps even all over the country.

The bilberry's small, inconspicuous flowers appear in April, and by late summer the shrubs carry dark-blue berries, known as fraughans, a likely anglicisation of the plant's Irish name, which are rich in vitamin C and antioxidants. Collecting the fraughans was once an important date in the calendar. The last Sunday in July was known as Fraughan (or Frochan) Sunday, or Heather-Berry Sunday, and on that day whole communities would have taken to the hills or into the bog to pick bilberries. In some areas, this activity was combined with courting; in County Kilkenny, young girls would bake a Fraughan Cake and present it to their chosen partner. And this is how bilberries were,

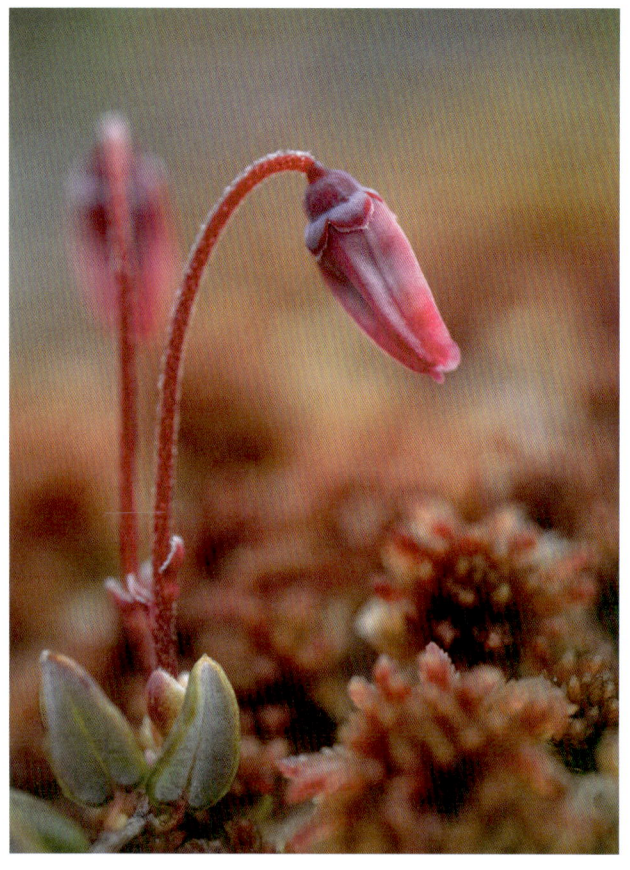

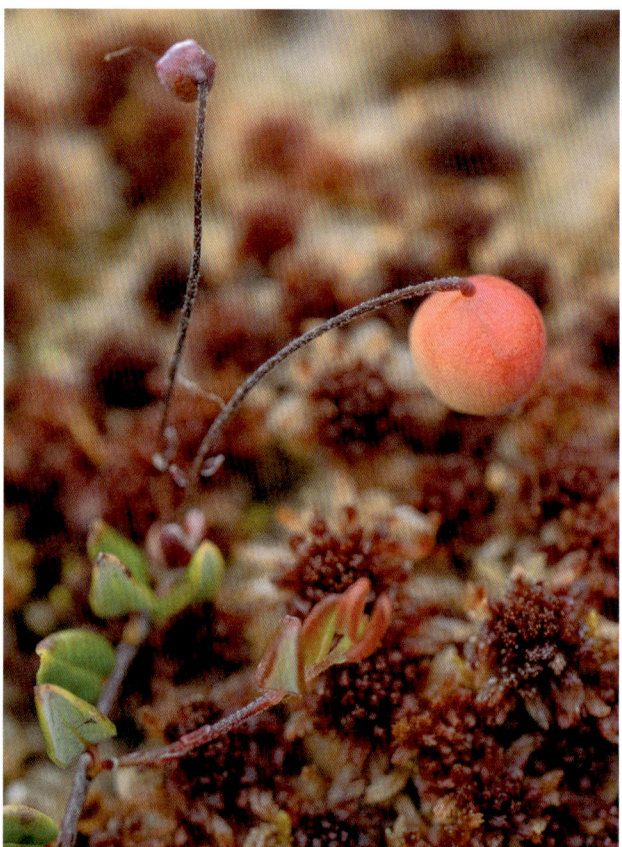

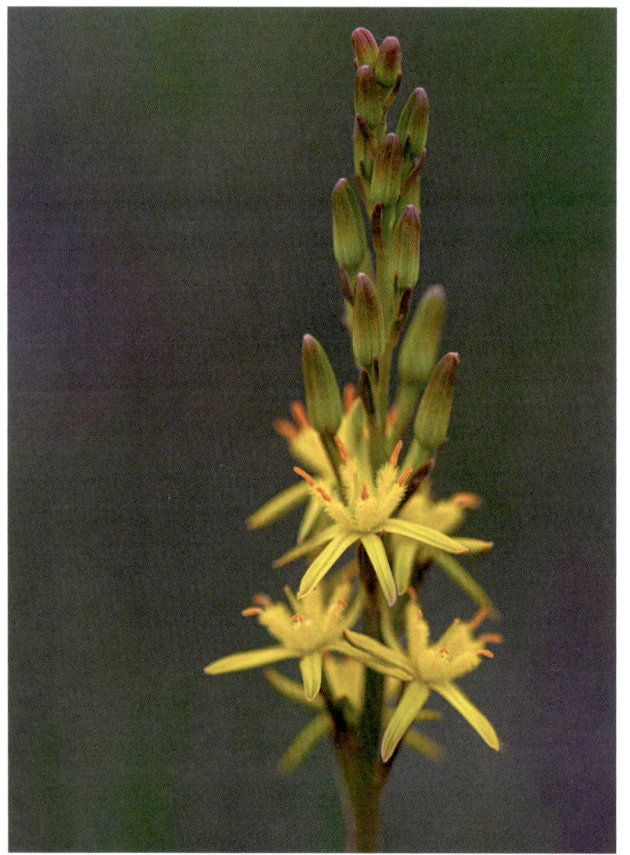

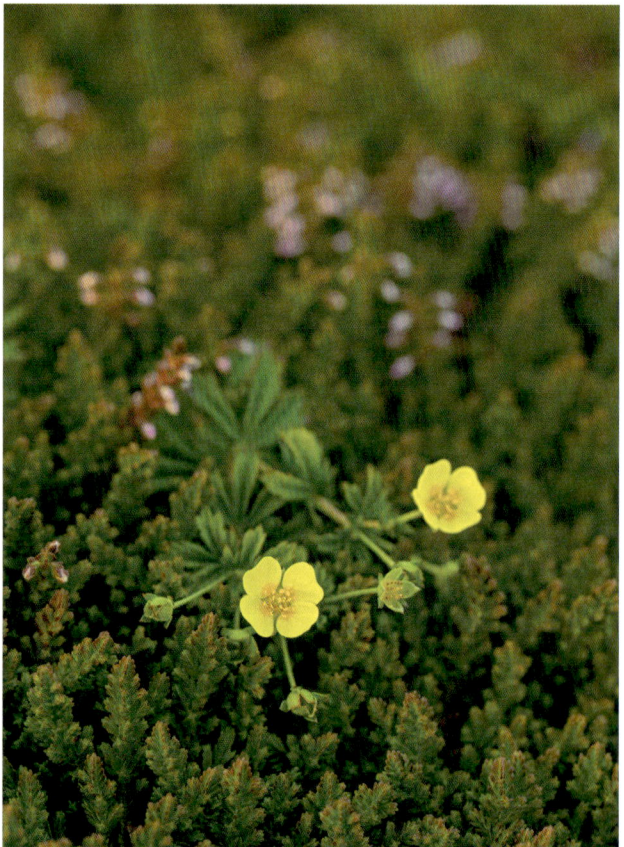

and to some extent still are, mostly used, in baked goods like cakes, pies, tarts or crumbles, because straight from the shrub they taste rather bitter and a bit of sugar is needed to make them palatable.

Another berry of the peatlands, these days mostly associated with Christmas dinner, is the cranberry. Its other names – moss-berry, moor-berry and bog-berry – hint at its preferred habitat, the raised bogs of the midlands and Northern Ireland. The cranberry is a small, creeping plant featuring delicate winding stems that grow close to the ground and stay mostly hidden beneath the carpet of sphagnum. Only the cranberry's bright-pink flowers get lifted above the mosses and in reach of the pollinators. During the summer, the flowers grow into white berries which over the course of a few weeks ripen into bright-red fruits with their characteristic bitter-sweet taste. Just like the bilberry, the cranberry carries anti-inflammatory properties and was once used widely for food as well as for medicinal purposes.

The cowberry, also known as lingonberry, is a close relative of the cranberry but rare and only known from a few locations in Ireland's north and east. It can be identified by its leaves, which are considerably bigger than those of the cranberry.

The crowberry also has a limited distribution in this country and grows mainly in the northern counties. This represents the southernmost expansion of the plant, which is widespread in Scandinavia and North America. Visually it is reminiscent of heather and in autumn carries dark blue berries, which have been a vital food staple in countries of the boreal zone.

Even rarer than the crowberry is the cloudberry, a relative of the bramble that produces blackberry-like, bright-orange fruits. This plant is only known from one location in the Sperrin Mountains in County Tyrone, where it was first described in 1826. Back then, its distribution was evaluated as 'plentiful', but shortly afterwards the cloudberry disappeared from any records. Only in 1892 did two field botanists discover two small patches of the plant after long hours of dedicated searching. Like the crowberry, the cloudberry is a typical inhabitant of northern countries, and Ireland seems to be at the southernmost border of its distribution zone. It is theorised that these plants are remnants of the post-glacial era when temperatures in Ireland were much lower. In 2003, only nineteen shoots of the plant were found. Because no flowers were ever discovered, it was assumed that these plants consisted of a single, most likely male, clone, unable to flower and produce seeds. In 2021, however, researchers from the National Botanic Gardens and the Northern Ireland Environment Agency spotted flowers on the plants, a discovery that might very much change the outlook for this lone representative of the cloudberry in Ireland.

Wildflowers

One of the most common and distinctive peatland wildflowers, thriving on both raised and blanket bogs, is the bog asphodel. Just like the heathers, this plant lives in a relationship with a mycorrhizal fungus to collect vital nutrients. Its characteristic spikes, laden with starry golden flowers, appear in June and can form extensive colonies. Where they do, the bog surface is transformed into a mesmerising flower meadow.

In some parts of the country, the crushed spikes and flowers were used as a yellow hair dye, leading

Opposite top left: Cranberry flower
Opposite top right: Cranberry fruit
Opposite bottom left: Bog asphodel
Opposite bottom right: Tormentil

> "
>
> ## THERE WAS AN OLD BELIEF THAT BROKEN BONES IN SHEEP AND OTHER DOMESTIC ANIMALS WERE CAUSED BY INGESTING BOG ASPHODEL
>
> "

to the plant's local name: maiden hair. In autumn, the spikes turn a vibrant orange and contribute a great deal to the warm autumnal colour scheme of the bog.

While bog asphodel is pretty to look at, it has a sinister reputation among livestock farmers. The second part of the plant's scientific name, *ossifragum*, translates into 'bonebreaker' and documents an old belief that broken bones in sheep and other domestic animals grazing on the bog were caused by ingesting bog asphodel. As it turns out, the flower indeed contains a chemical which can have an adverse effect on bone strength – and unfortunately, it doesn't stop there. Bog asphodel also contains a substance that can cause serious kidney and liver damage to sheep, a condition known as 'elf-fire', as well as a photosensitivity disorder known as 'yellowses' or 'plochteach'. Symptoms start with the swelling of face and limbs, which causes the animal to rub its head against a hard surface up to the point where blood is drawn, and prolonged ingestion eventually leads to the death of the animal. The broken bones, however, are more likely to be caused by bad nutrition from grazing on the bog, primarily a lack of calcium, which can lead to brittle bones.

Flowering at the same time as the bog asphodel and often growing nearby are the relatively common heath spotted-orchid and the rarer bog orchid. While the first, with its pink and white colour scheme, is easy to spot, the latter is a small and inconspicuous flower, its pale, yellow-green colour perfectly blending into the background of mosses, sedges and grasses. Other orchids that can be found on bogs are the lesser twayblade, which prefers ling-dominated blanket bogs, and the marsh orchids. The latter comes in a variety of species, sub-species and variations that can drive even seasoned botanists into confusion.

A flower that is as delicate and gracile as the bog asphodel but flowering earlier in the year is the bog bean. This wildflower can often be found in shallow pools and ponds, and its unmistakable star-shaped flowers are like no others in Ireland. Elegant white or slightly pinkish petals are covered in tiny erect hair, giving the flower a fluffy appearance, and surround a vivid yellow, pollen-covered stamen.

The bog bean's common name refers to its leaves, which according to 16th century herbalist John Gerard look very much like the ones of the garden bean. The two plants are, however, not related at all. The roots of the bog bean, also known as 'bog bine' or 'bog hop', were traditionally collected in spring and then boiled and made into a tonic 'to clean the blood', allegedly setting the consumer up to stay in good health for the whole year. In some localities, the bog bean was also used as a replacement for hop to flavour beer.

Another wildflower that was given its name because of its resemblance to an unrelated species

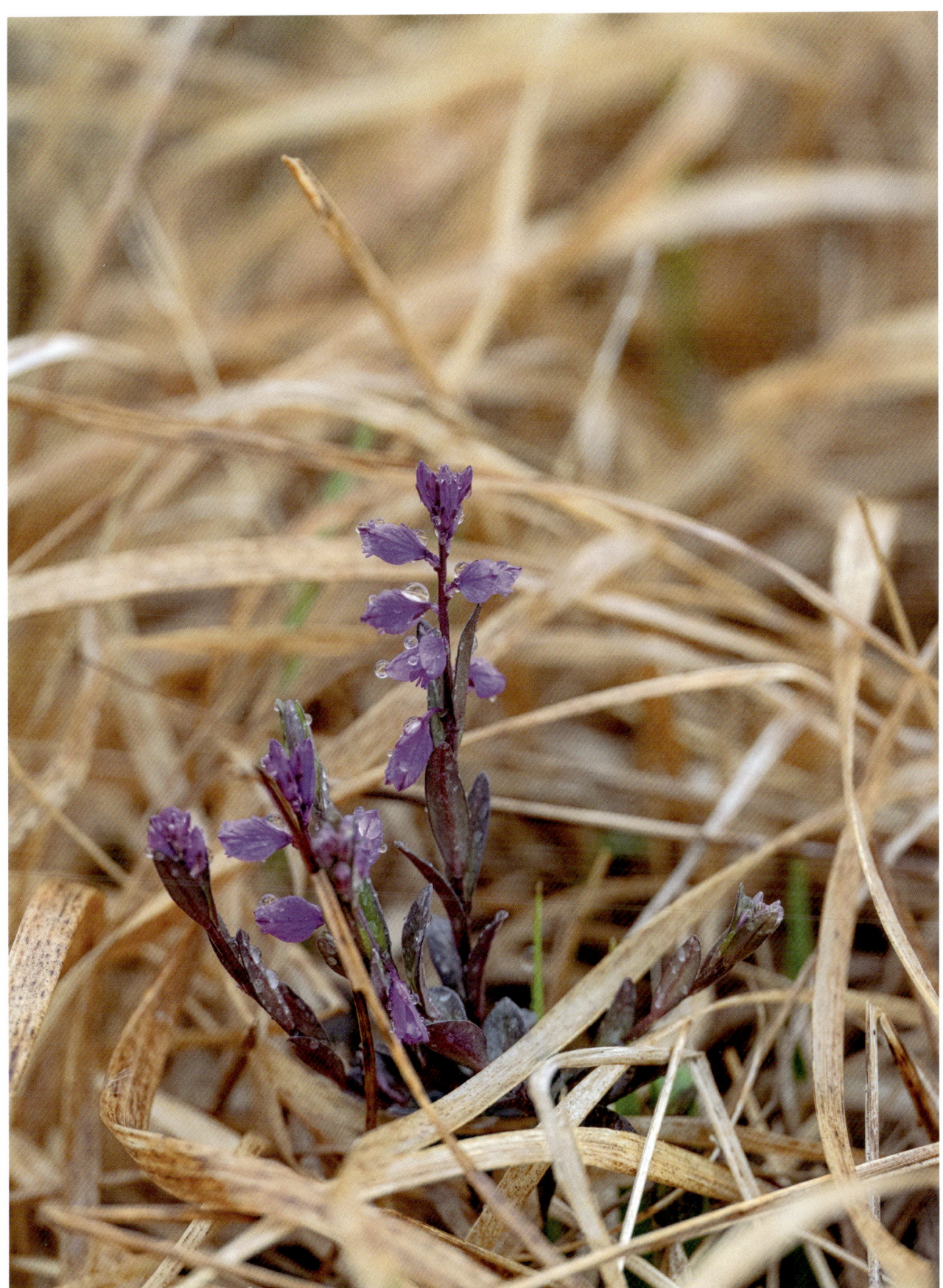

A pink version of the heath milkwort

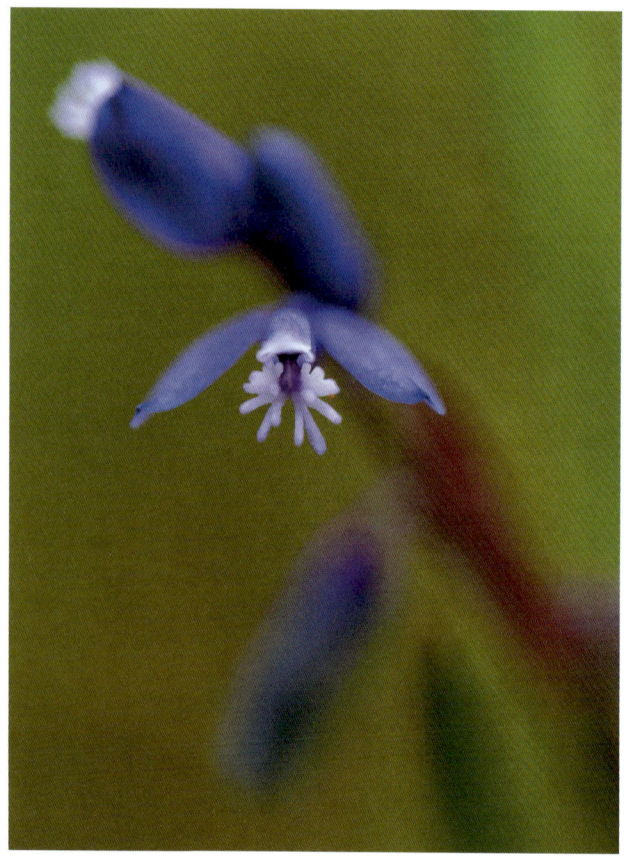

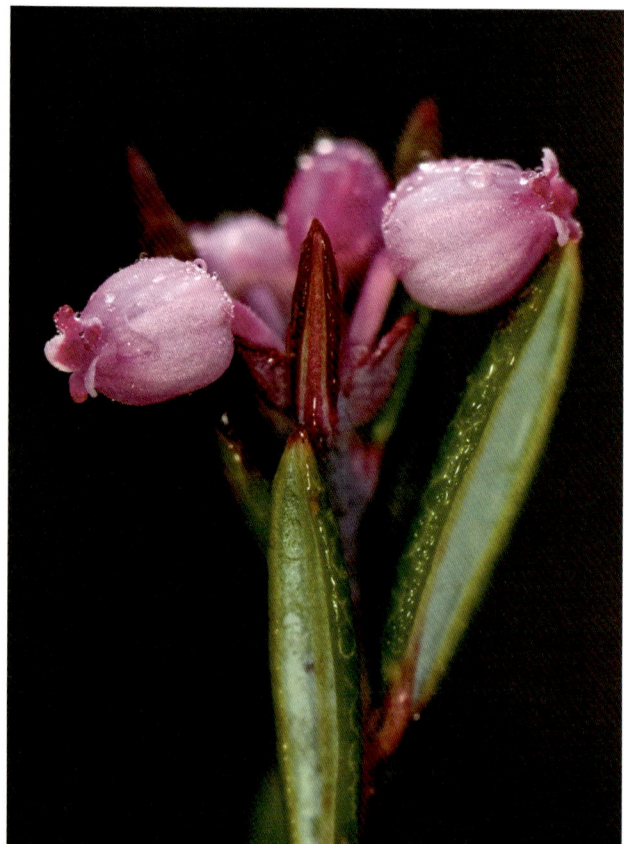

is the bog rosemary, a typical flower of the raised bogs most common in the midlands and missing from all blanket bogs. Unlike its culinary namesake, bog rosemary is not only unpalatable but actually poisonous. The plant, which is botanically speaking a shrub and a member of the heather family, contains andromedotoxin, also known as grayanotoxin, which affects the central nervous system and can lead to cardiac arrhythmia and paralysis. Depending on the amount consumed, ingestion can be fatal. The toxin is present in all parts of the plant, but the highest concentrations have been found in the leaves during the winter months, a very effective defence mechanism to keep grazers at bay during the time of year when little food is available.

Milkwort, tormentil and lousewort are a characteristic and colourful trio of wildflowers of blanket bogs and heaths. Common milkwort and heath milkwort are almost identical plants with small but strikingly blue flowers (paler versions in shades from pink to purple are also known), which differ only in the arrangement of their leaves. Heath milkwort has its leaves arranged in opposite pairs; common milkwort shows an alternating pattern. The name 'milkwort' was given to the plant because it was thought cattle feeding on this plant produced more milk, a belief so deep rooted that the plant was also given to nursing mothers.

Tormentil forms creeping patches close to the ground and produces four-petalled, bright-yellow flowers. It was widely used to treat digestive problems, sleep disorders and inflammations, which is reflected in the various names of the plant. Its common English name is thought to be derived from the Latin *tormentum*, referring to the stomach pain the plant relieves. Its Irish name, *néalfartach*, is a combination of the words *néal* and *fartach*, meaning 'gloom' and 'hurt' respectively. In County

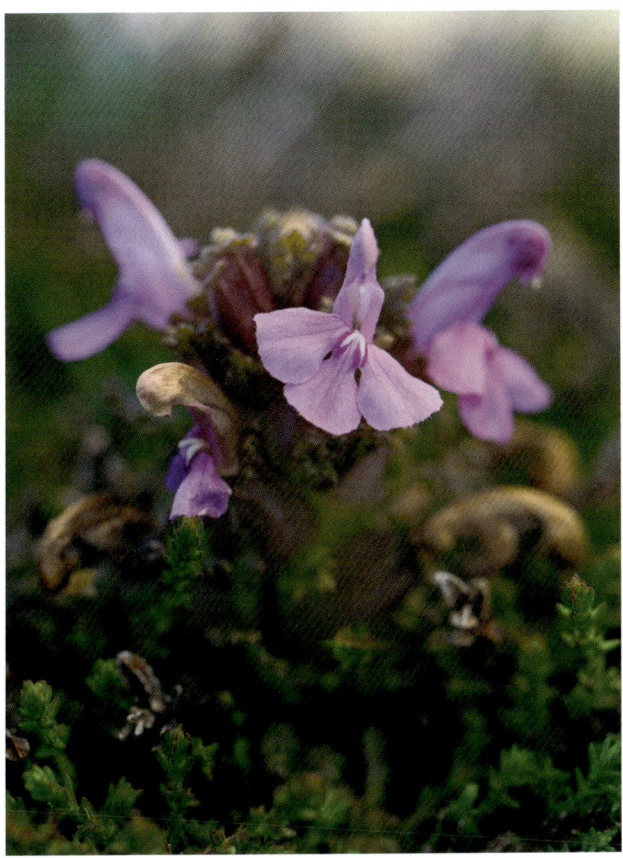

Opposite left: A deep blue is the most common colour for the flower of the heath milkwort
Opposite right: Bog rosemary
Left: Lousewort

Cork, it was known as *lus an chodlata*, 'herb for sleep', and *neal codladh*, 'snooze' or 'wink of sleep'. Research has indeed shown that tormentil roots contain high amounts of tannins, which have antimicrobial and astringent properties and are therefore well suited to treat inflammations and offer relief from digestive problems caused by bacteria.

The last in the trio is lousewort, which got its name from the belief that it gave lice to livestock. Lousewort is a semi-parasitic plant that taps into the roots of neighbours to increase its own nutrient intake. This might have been the reason behind its name – after all, the louse is one of the most feared parasites around. When in bloom, lousewort carries beautiful pinkish flowers, and the whole plant is reminiscent of a shrunken orchid. Slightly less widespread but more impressive is the marsh lousewort, which grows to almost twice the size of its relative and carries stronger-coloured flowers.

Lichen

Lichen are enigmatic entities, unique among all plants and animals. Each lichen is a close-knit partnership of fungi, algae and cyanobacteria. The fungi provide structure, protection and moisture, while the algae or the cyanobacteria, or sometimes both, are responsible for energy production through photosynthesis. This setup allows them to thrive in challenging environments where other life forms wouldn't be able to survive.

The most typical peatland lichen, which equally appear on raised bogs, blanket bogs and heaths, are the cladonia species, namely the aptly christened matchstick lichen (*Cladonia floerkeana*), also known as 'the devil's matchsticks', the delicate pixie cup lichen (*Cladonia pyxidata*) and the intricate antler horn (*Cladonia uncialis*) and bearded lichen (*Cladonia portentosa*).

Depending on the location and environmental

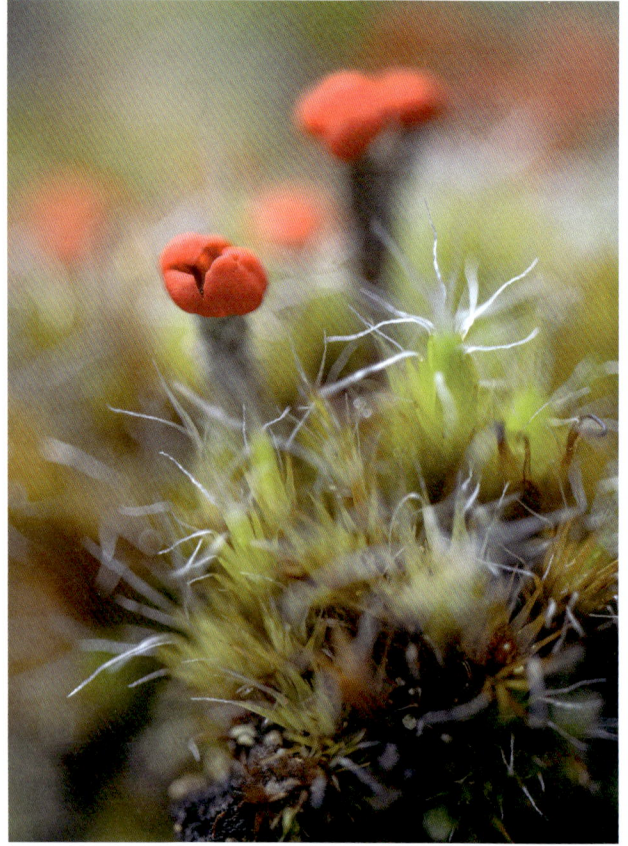

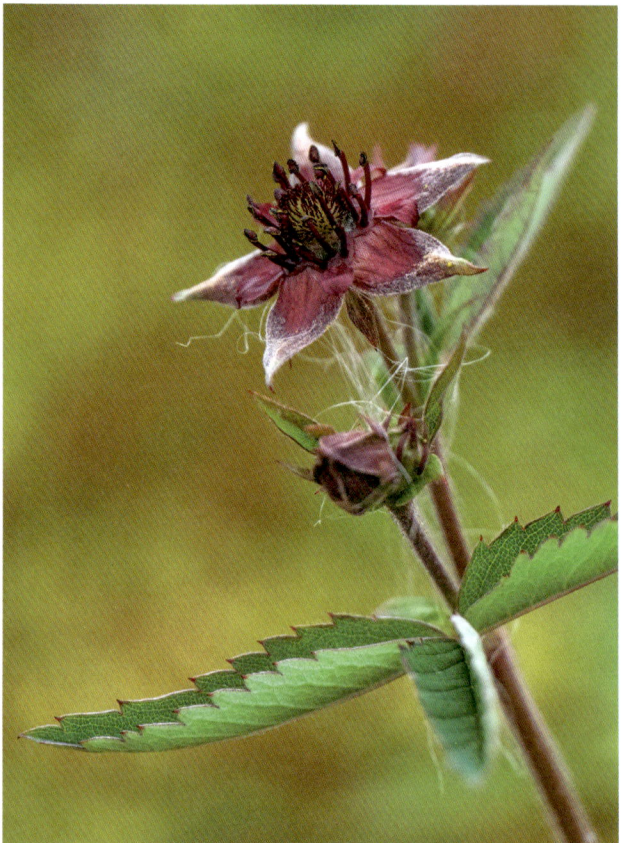

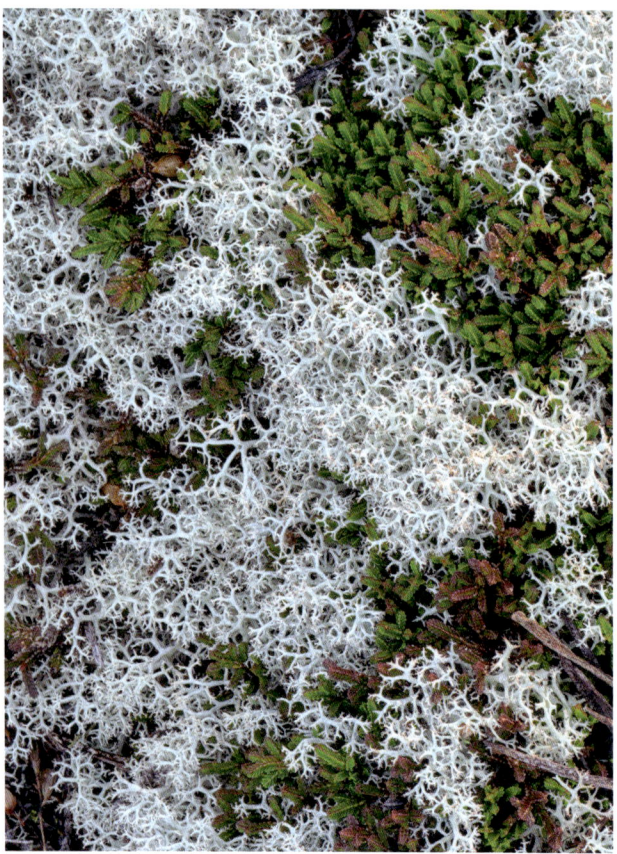

Top left: Devil's matchstick lichen
Top right: Marsh cinquefoil
Left: *Sphaerophorus globosus* and heather

conditions, some blanket bogs and heaths along the west coast of Ireland also host a variety of other lichen species. Connemara in particular, where bare granite rock breaks through the bog surface and the landscape is scattered with massive erratic boulders, is very inviting to lichen. Leafy species like the blue-grey *Hypogymnia physodes* and the silver-grey *Parmelia saxatilis* are common. The latter was historically used as wool dye, giving a warm, reddish-brown colour to jumpers and socks. Some unusual individuals are also present: the beard-like *Bryoria fuscescens* and the distinctive green-grey and black-dotted *Mycoblastus sanguinarius* or bloody-heart lichen are typical woodland species and likely survivors from the post-glacial pine forests that traded in trees for rock in order to survive.

Fen Wildflowers

The flora of fens, particularly calcareous fens, is considerably more varied than that of bogs and heaths. The environment is generally more hospitable, and the all-important nutrients are more readily available. While poor fens host primarily plants that can also be found on bogs and heaths – a variety of mosses, grasses, rushes and sedges, carnivorous species and bog bean – as well as some species that can tolerate slightly acidic conditions like the cuckoo flower, marsh violet and lesser spearwort, calcareous or rich fens support a much wider range of plant life including many common wet grassland species like angelica, hemp-agrimony, wild valerian and meadowsweet. The last is the foodplant of a variety of moth larvae, including the emperor moth, grey

pug, satellite and mottled beauty. Humans also cherished the meadowsweet, and it was long recognised as a potent painkiller and anti-inflammatory medicine. In the 19th century, researchers isolated salicylic acid from the plant, which became the active ingredient in aspirin.

Similarly important was water mint, which carries the same properties as other mint species and was once widely used for digestive problems and headaches. In the Middle Ages, water mint was also strewn on the floor of dwellings to freshen the air, and in County Kerry, it was smoked as a tobacco replacement. A variety of insects like the small tortoiseshell, peacock and comma butterfly are very fond of the water mint's nectar.

Another common wildflower of fens and other wetlands is the cuckooflower. With its mostly white flowers, which can at times show a beautiful purple or pink hue, it is the exclusive foodplant of the orange-tip butterfly larvae. It was named for a flowering period that coincides with the arrival of the cuckoo, as 16th century herbalist John Gerard describes: 'These flower for the most part in April and May when the Cuckoo begins to sing her pleasant notes without stammering.'

The devil's bit scabious is the sole larval foodplant of the endangered marsh fritillary butterfly, and a rare bee, *Andrena marginata*, also uses the flower almost exclusively to collect pollen as food for its offspring. Marsh fritillary larvae hatch from their eggs in mid-June, and then as a team spin a web around the lower part of the devil's bit scabious plant they were born on. Over the following weeks, this web is extended to house the

"

THE ROOTS OF THE BOG BEAN WERE BOILED AND MADE INTO A TONIC 'TO CLEAN THE BLOOD'

"

growing caterpillars. Eventually, neighbouring plants are included into the caterpillar mansion, and by September, the web has grown to a considerable size. The black caterpillars overwinter in this web and become active again the following March for one last big communal feast before they disperse to feed on their own and to pupate around the end of April.

Devil's bit scabious got its unusual name because of the shape of its roots, which look like someone has bitten them off. According to legend, this someone was none other than the devil, who was annoyed about the medicinal properties the plant had to offer. Devil's bit scabious was used to treat a variety of skin conditions including symptoms of syphilis and plague, and the effects of the sap of lesser spearwort and other poisonous plants. Lesser spearwort is a member of the buttercup family. It showcases flowers that are very similar to the widespread creeping and meadow buttercup, and can often be found growing at the edge of calcareous fen lakes. Unlike the buttercups, lesser spearwort produces a bitter and acrid sap that causes severe blisters when it comes in contact with skin.

One of the rarest fen wildflowers is the yellow-flowered marsh saxifrage, which has been recorded from only two sites in Ireland: one in County Mayo and another in County Antrim. Surprisingly, neither of these sites is within a fen, and the yellow-flowered marsh saxifrage, although considered a fen species, currently only thrives in mineral-rich flushes within blanket bog areas.

The complete list of fen wildflowers is long and includes widespread species like yellow flag or flag iris, with its intricate yellow flowers, which was once widely used to treat coughs, stomach ailments and insect bites. Marsh marigold, also known as kingcup, is another poisonous member of the buttercup family named after the Virgin Mary. It used to be popular at Easter celebrations, and in some parts of the country where the plant was known as mayflower, it was placed on the threshold and window ledges of homes on the last day of April – May Eve – to ward off evil spirits. Rarer than these two species is the marsh cinquefoil, with its unmistakable marron-coloured flowers. Its distribution has declined considerably due to the drainage of its habitat, and today it is mostly to be found in drains and ditches.

A variety of orchids are also at home in fens, including the fly orchid, marsh helleborine, common spotted orchid, butterfly orchid and various marsh orchids.

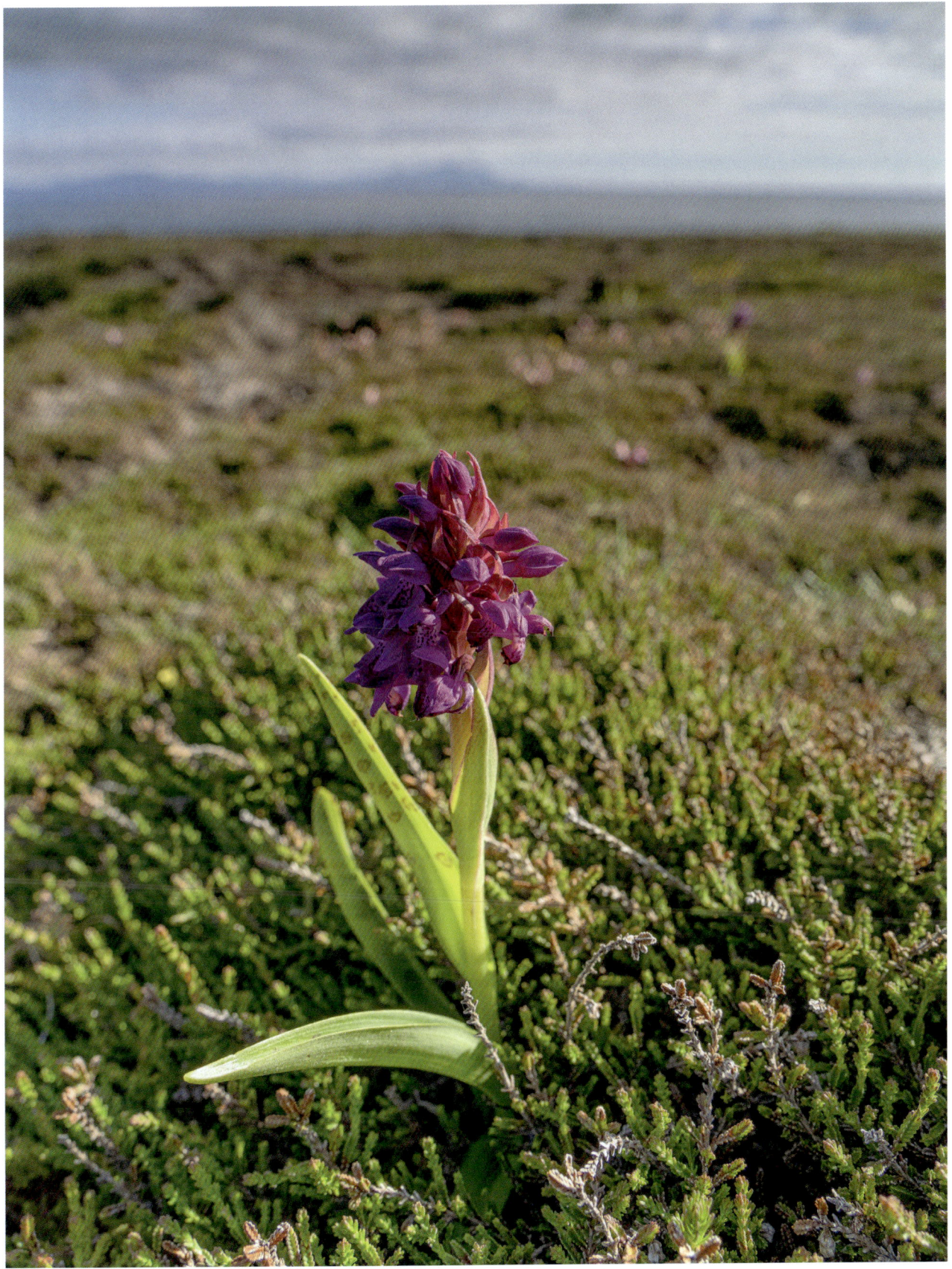

Early marsh orchid

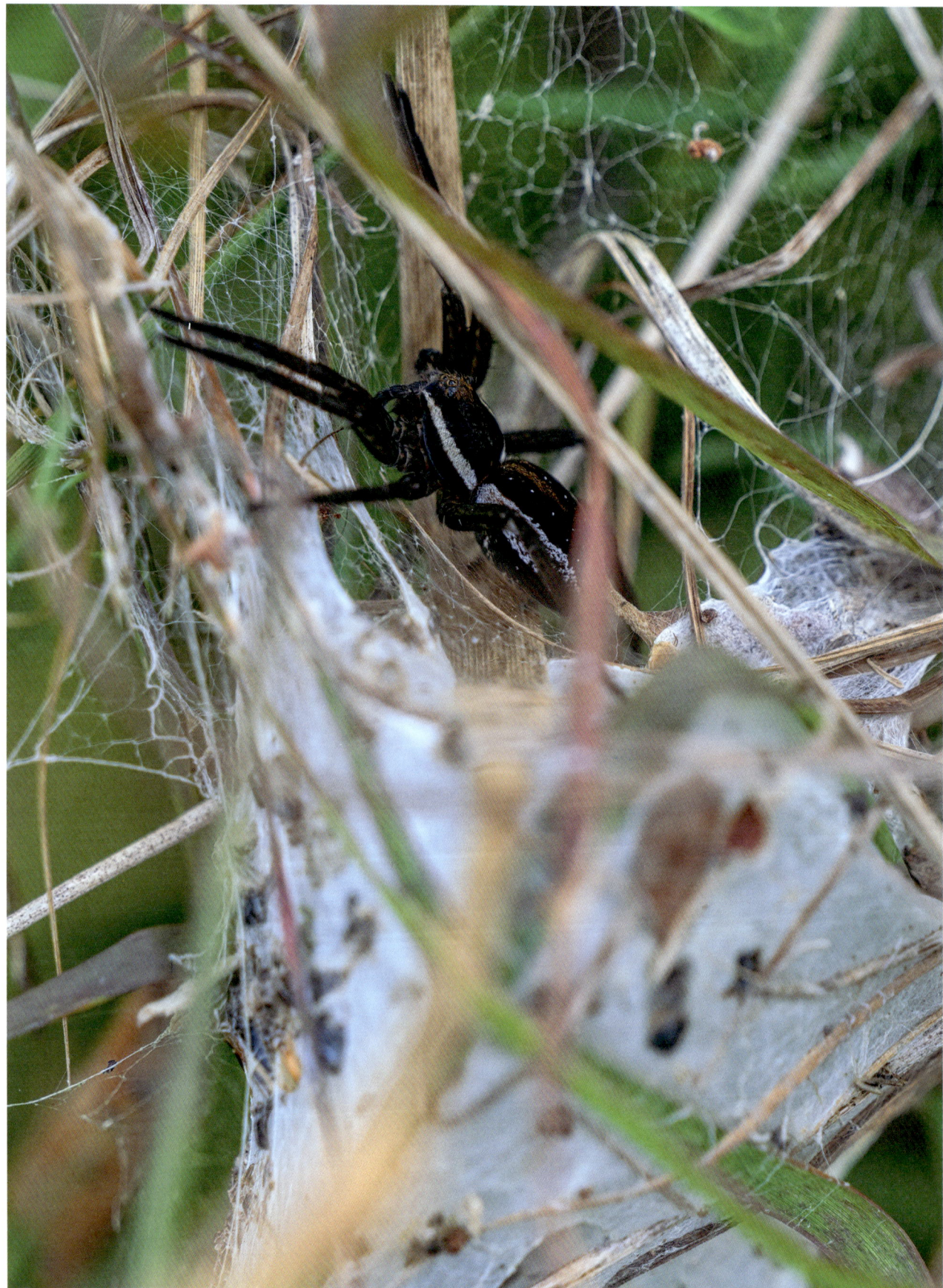

Raft spider

"

THE RAFT SPIDER RACES OVER THE WATER TOWARDS ITS PREY, WHICH GAVE RISE TO ITS NICKNAME 'THE JESUS SPIDER'

"

FAUNA

While quite a few plants have adapted to life in the peatlands and exclusively grow on bogs and heaths, most animals encountered here can also be found in other habitats. Some visit the peatlands only to feed; some choose to raise their young here but also use neighbouring habitats to search for food; some come for the summer, others for the winter. Very few have made the bog and heath their permanent and only home. Most of those permanent residents are invertebrates, insects and other creepy-crawlies that spend their entire life cycle, from larvae to adult, in peatlands, often hidden deep in the vegetation.

Invertebrates

The most plentiful invertebrates on peatlands are flies, midges – including the infamous biting *Culicoides impuntatus* – and other small insects. A study conducted in County Mayo found 106 species of mites, represented by 18,000 individuals, and 64,000 springtails in only one square metre of blanket bog. These small animals form the bottom of the food chain and provide sustenance for bigger hunters like the raft spider, Ireland's largest and arguably most beautiful arachnid. The body of the female raft spider can measure up to two centimetres, and the whole animal can easily fill the palm of a hand. It is covered in brown, sometimes greenish, hair and features distinctive white stripes running along both sides of the body.

Raft spiders are semi-aquatic and prolific hunters. They lie in wait at the edge of ponds or on floating vegetation on bog pools, with two or more feet touching the water's surface. The moment the spider senses even the slightest ripple, it races over the water towards its prey, a behaviour that gave rise to its nickname 'the Jesus spider'. To perform this trick, the spider takes advantage of the water's surface tension and uses air bubbles trapped in its leg hair for additional buoyancy. If necessary, the spider can also dive after its prey. If the hunt was successful, they bring their prey – which can range from insects to snails, tadpoles and even small fish – back to the shore to eat.

As with most arachnids, courtship is a tricky affair for the male, which is a good bit smaller than the female. In springtime, males seek out females, approaching carefully, and when the female spider is agreeable, both perform a mating dance, bopping their bodies up and down. Then mating takes place, after which the suitor makes a run for his life. Raising the offspring is the responsibility of the female. She spins an egg sack, into which she lays several hundred eggs, and for about three weeks she carries this sack around, every once in a while dipping it in water to avoid the eggs drying out. Shortly before the eggs hatch, the female spider builds a nursery-net close to the water's edge, places the eggs in it, and stays close by for some ten days to protect the newly hatched spiderlings. Once big enough, the youngsters disperse to find a place to hibernate for the winter.

Ireland's only other aquatic spider, the water spider or diving bell spider, can be found in (but is not restricted to) peatlands, and it takes the

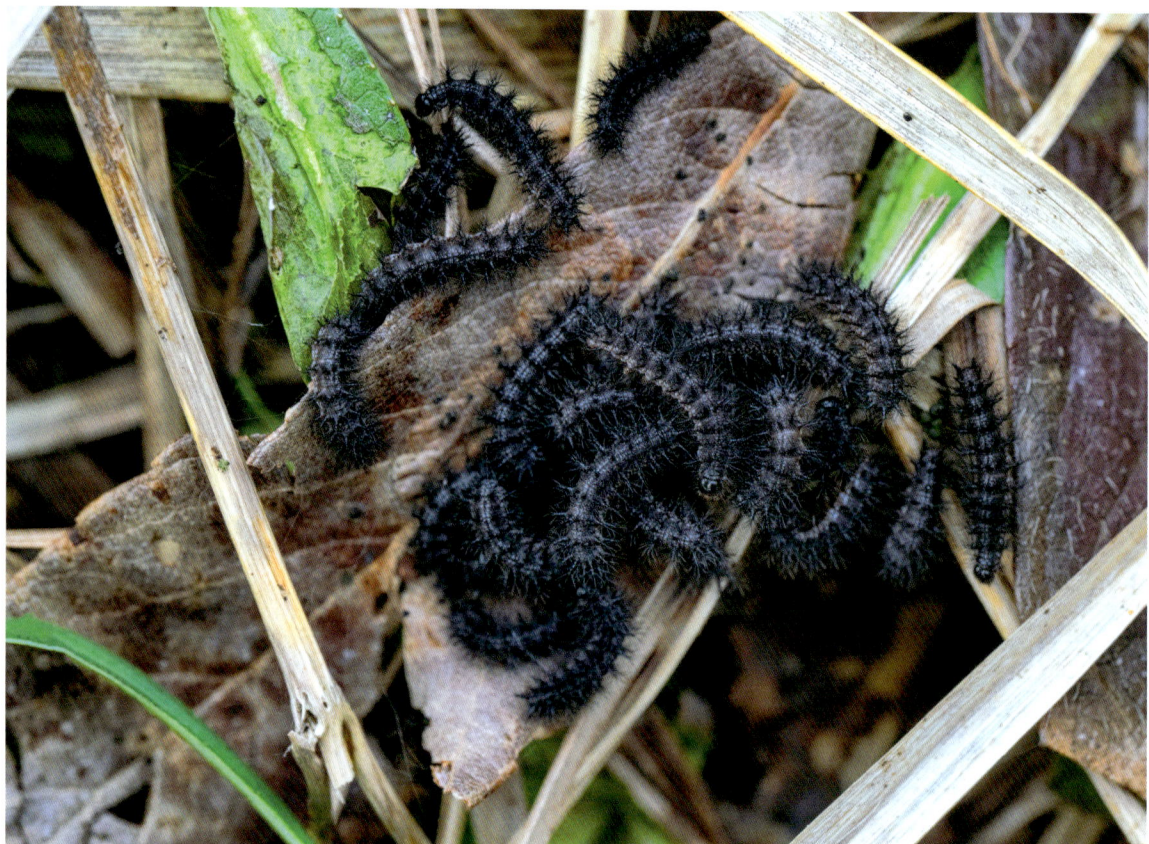

Marsh fritillary caterpillars

aquatic lifestyle to the extreme. This species spins a bell-shaped net beneath the water surface. It traps air in the fine hairs on its body and then makes several trips to the bell to transfer the trapped air into its new underwater home. The opening of the bell points downwards, which prevents air from escaping and offers the spider inside a good 360-degree view of any passing prey. The bell is anchored with several silken threads to nearby vegetation. The spider stays inside its diving bell, front legs sticking out to sense and catch the next meal. The prey is then consumed inside the diving bell.

Another typical peatland species is the large marsh grasshopper, which can reach a length of over three centimetres and comes in a vivid yellow-green or olive-brown colour with black knees and a characteristic red streak on its hind femur. It is the largest grasshopper species of the European mainland and unfortunately also the rarest; in Ireland, the insect is restricted to only a few sites in the southwest, west and midlands, where it resides in the wettest parts of raised and blanket bogs. Measuring up to its size, the large marsh grasshopper is also the loudest, its call sounding like the amplified ticking of an electric pasture fence.

The beetles of the Carubus family come in a variety of species, all of which are relatively large, measuring on average between two and three centimetres. The most common is *Carabus granulatus*, coloured in granulated bronze, green or black and living in upland peatlands and other wet environments. *Carabus clatratus* can make do in the bog or other places with waterlogged

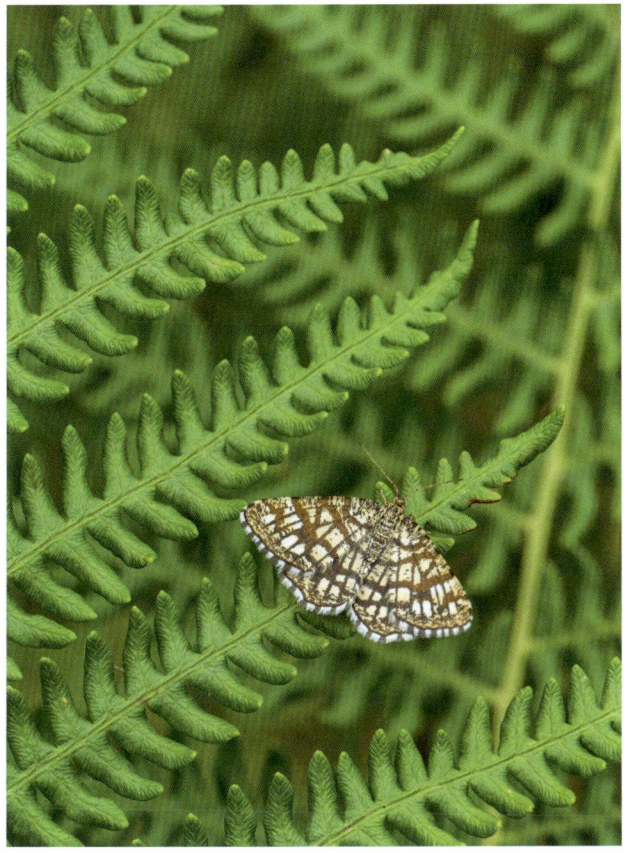

Latticed heath

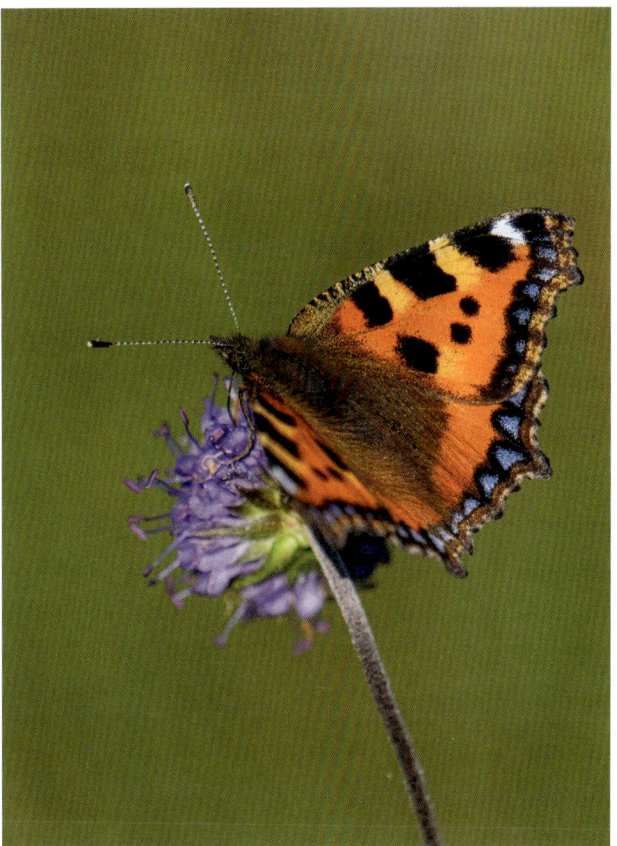

Small tortoiseshell on devil's bit scabious

ground and comes in a metallic bronze, green or black colour. *Carabus arvensis* is one of the smaller species, which showcases a granulated bronze or green colour and lives mostly in drier parts of heaths and blanket bogs. *Carabus glabratus* is one of the biggest species, with individuals measuring more than three centimetres. This dull, metallic-black coloured beetle prefers heather patches in uplands and is most common in Ireland's southwest.

Other insects that are regularly encountered in (but not limited to) peatland habitats are dragonflies and damselflies like the four-spotted chaser, black darter, emerald damselfly and large red damselfly. The same goes for butterflies and moths, including the green hairstreak (which feeds on gorse among other foodplants), meadow brown and small tortoiseshell (both feeding on a variety of grasses), emperor moth (which feeds on heather and meadowsweet) and large yellow underwing (which is not very picky and feeds on various herbs and grasses).

There is one exception though: the large heath. This medium-sized butterfly – which features tawny upper wings with dark spots, and grey to rusty-brown underwings with an irregular white streak and one to six ringed eyespots – is a peatland specialist. Its caterpillars, which are green with a darker-green dorsal line bordered by a white stripe, feast exclusively on hare's-tail cotton grass, and the adult butterfly only drinks nectar from cross-leaved heath. Because of this specialised diet, the large heath needs intact peatlands, where both plants are abundant, to survive.

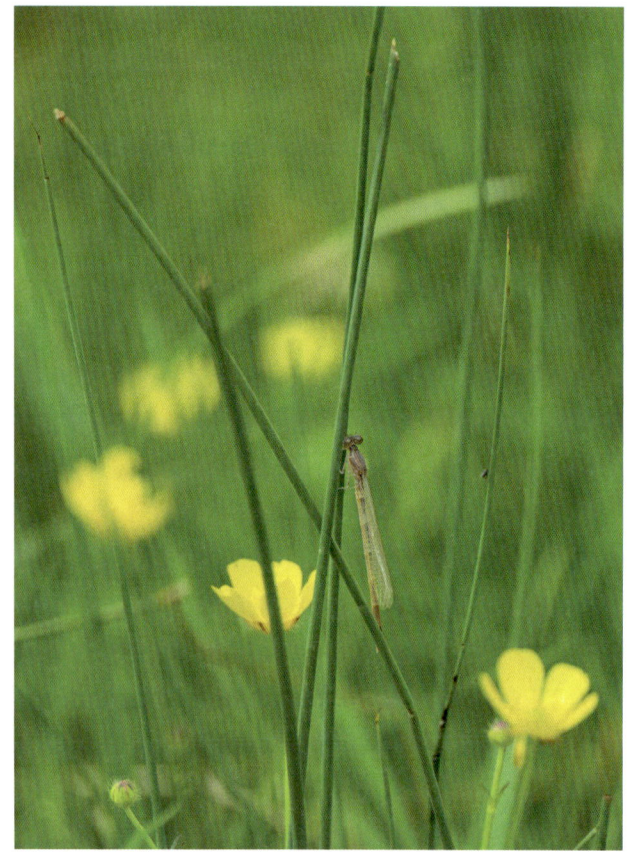

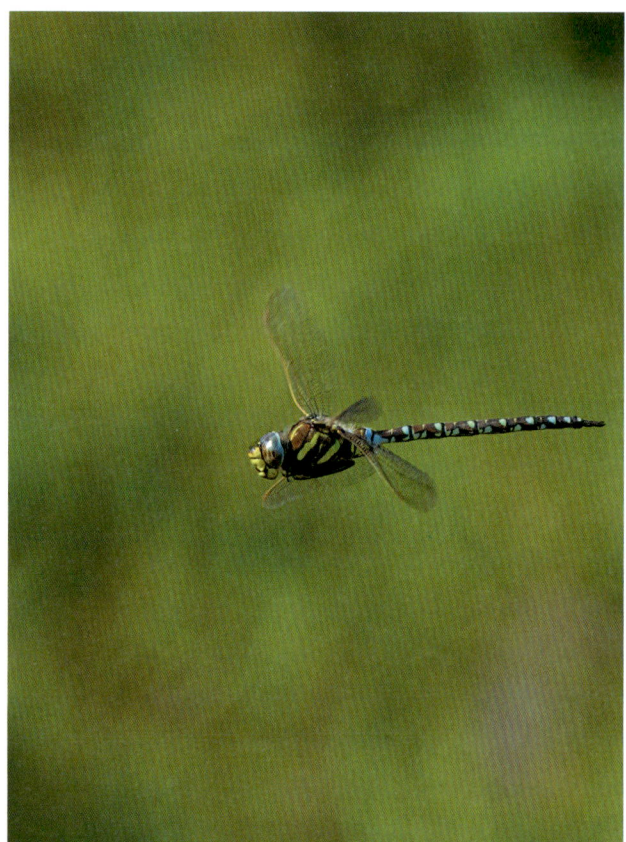

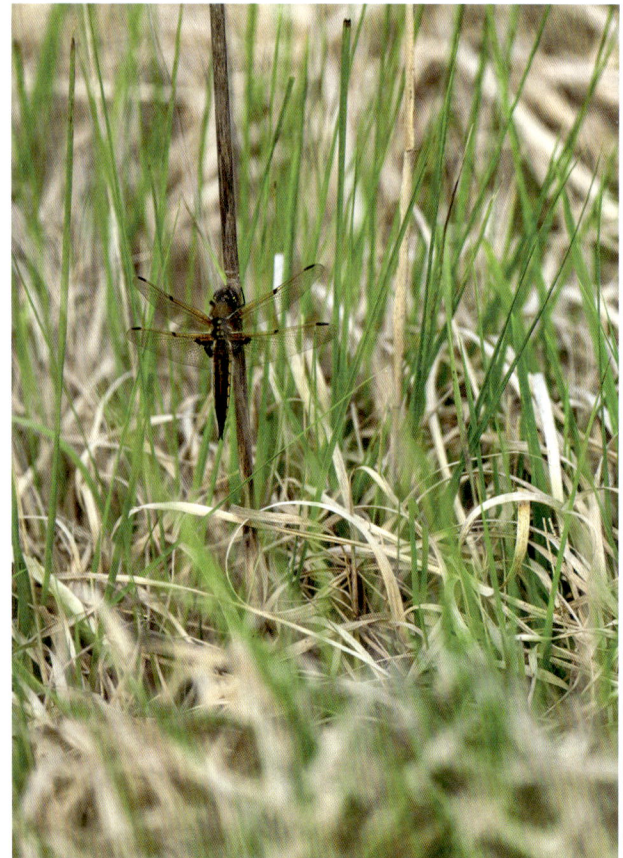

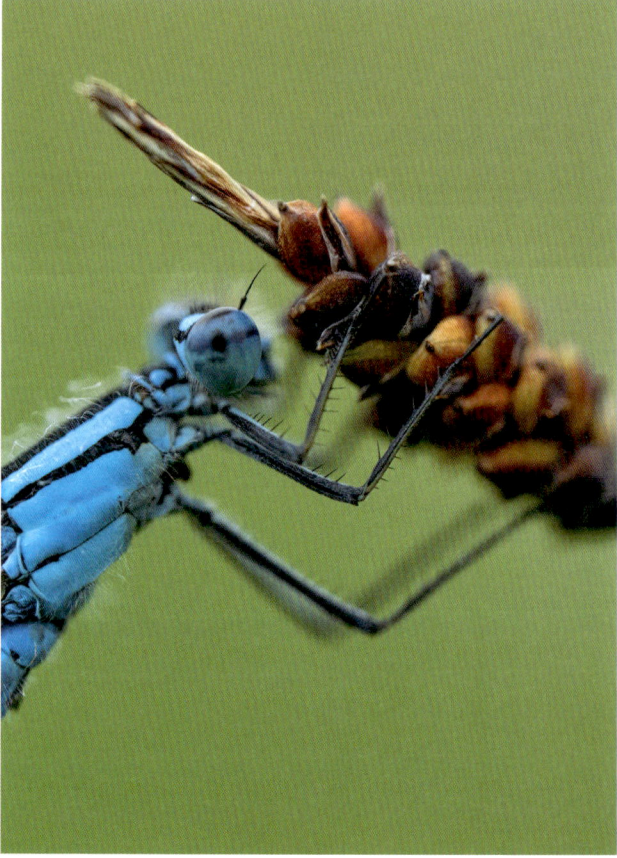

Opposite top left: Teneral (just hatched) blue-tailed damselfly
Opposite top right: Common hawker
Opposite bottom left: Four spotted chaser
Opposite bottom right: Common blue damselfly

Amphibians and Reptiles

It should come as no surprise that amphibians feel very comfortable in the wet peatland environment. Ireland's most widespread amphibian, the common frog, appears anywhere there are some damp spots to be found; in forests, meadows, and of course fens, heaths and bogs. The availability of pools, ponds and lakes only becomes important in springtime, when males and females get together to mate. In this delicate procedure, the smaller male clambers on the female's back and holds on for dear life using the so-called nuptial-pats on his front feet. This part of the mating procedure can become very chaotic and even dangerous for the female when several love-sick males try to secure a piggy-back ride at the same time. Once one male has emerged victorious and the pair is comfortable, eggs and sperm are released at the same time and fertilisation takes place. One pair can produce up to 2,000 eggs at a time. The number of eggs must be this high as they provide a nutritious meal for a range of animals, and many eggs get devoured before the tadpoles can even hatch. The lucky tadpoles that appear after about four weeks have at least a chance to escape their predators, and with some additional luck they grow into fully formed frogs in another three to four weeks.

The grown-up frogs are perfectly adapted to their environment. Their bulging eyes have a 180-degree view each and can scan the surface while the body stays hidden underwater or under a carpet of sphagnum moss. To make the camouflage perfect, frogs can adapt the colour of their skin to match their environment. This is done through specialised skin pigment cells called chromatophores. These come in different varieties; some are filled with darker pigments, others with lighter pigments, and the frog can adapt its colour by expanding or contracting these cells. The camouflage not only helps the frog stay hidden from predators but also gives it an advantage sneaking up on its own prey – flies, moths, snails, slugs and worms. The frog's secret weapon when hunting is its tongue. It measures around one-third of its body length and produces a very sticky saliva when needed. Once a target is acquired, the frog shoots out this tongue, which hits the prey with a force of about 12 Gs, approximately the same force astronauts experience during a rocket launch. This impact stuns the victim, and the sticky saliva makes sure the meal can be reined safely into the frog's mouth. All of this takes no longer than 0.07 seconds; five times faster than the blink of a human eye.

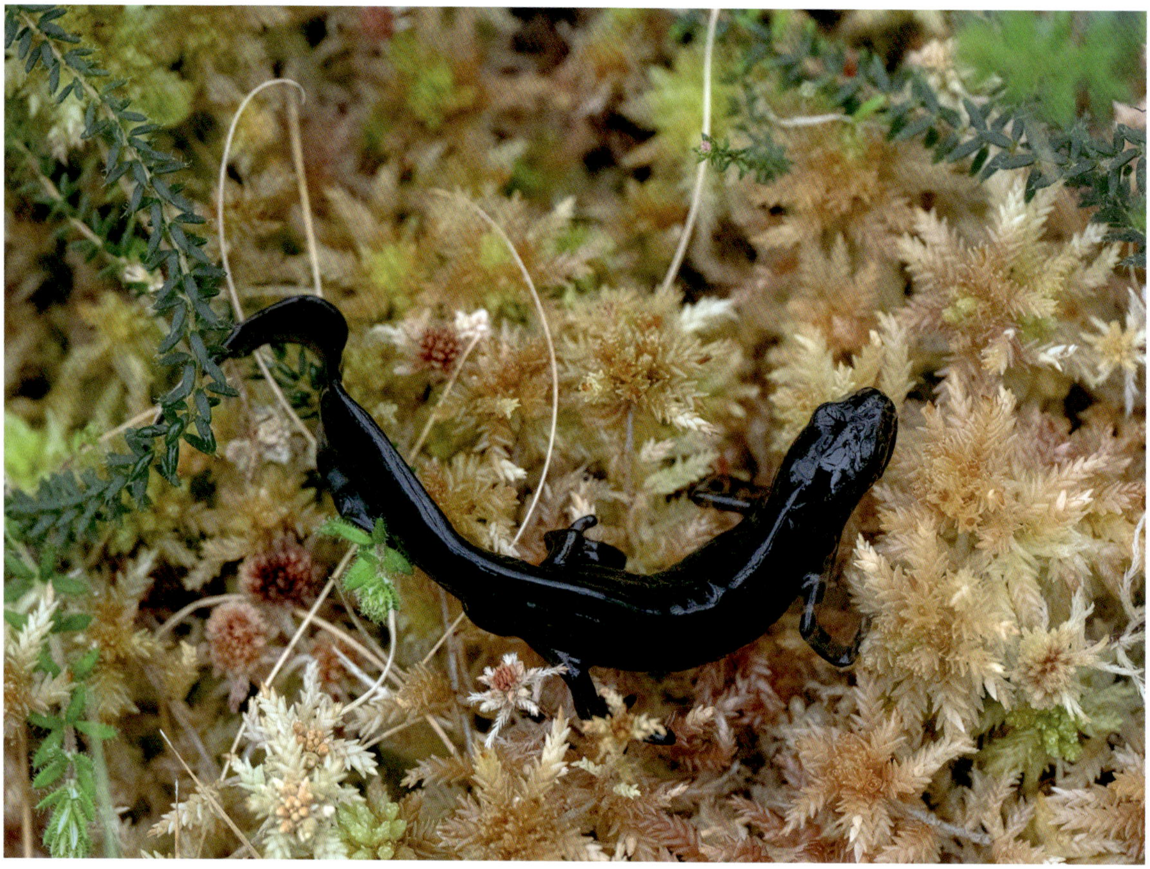

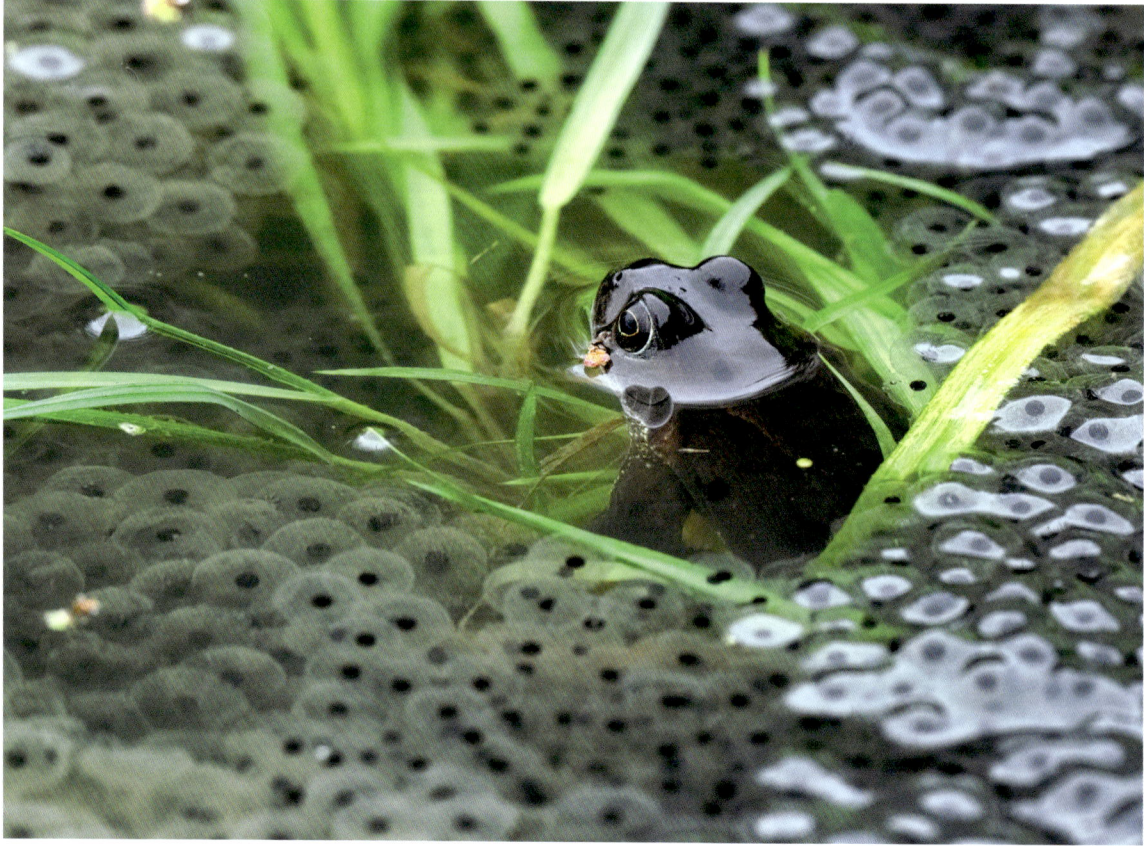

THE COMMON FROG APPEARS ANYWHERE THERE ARE SOME DAMP SPOTS TO BE FOUND; IN FORESTS, MEADOWS, AND OF COURSE FENS, HEATHS AND BOGS

The smooth newt, just like the common frog, spends most of its life on land and only ventures into the water to breed. Unlike the common frog, however, the smooth newt uses its chromatophores not only for camouflage but also to attract a mating partner. In the breeding season the animal's orange belly becomes more vibrant and the black spots more prominent. In addition, the male produces a dorsal crest and expands the webbing on its hind feet.

The smooth newt's mating ritual is a bit more civilised than that of the common frog. The male performs a mating dance to attract a female, and once he has successfully seduced a partner he guides her over a previously deposited sperm packet. The female picks up the sperm packet and the eggs are fertilised internally and then laid in a suitable spot. Out of them, tadpoles, known as 'efts', hatch and develop into adult smooth newts, which stay in their birthplace until late summer when they scatter to find a safe spot to hibernate.

Like the common frog and smooth newt, Ireland's only native reptile, the common lizard, is not limited to peatland habitats but can be encoun-tered there sunbathing on hummocks and rocks. In northern countries, this species is also known as the viviparous lizard, arising from the fact that it apparently gives birth to live young, unlike other reptiles who lay eggs. Perceptions, however, can be deceiving. Eventually it was discovered that the viviparous lizard bears its offspring through eggs just like any other reptile but has adapted to the colder climate in northern Europe. To keep the eggs warm and ensure their proper development, the female lizard keeps them inside her body up to the time when the young lizards hatch.

Mammals

While Ireland's largest mammal, the red deer, can often be found in upland blanket bog areas, it is not really a peatland species. The red deer originated as an animal of forests and grass-lands and has been present in Ireland for over 5,000 years. Deforestation and the spread of peatlands forced the animals to adapt, but the disappearance of their original habitat and ruthless hunting brought the population close to extinc-tion. In the early 20th century, only sixty animals

Opposite top: Smooth newt
Opposite bottom: Common frog

"

THE IRISH HARE HAS BEEN CONTINUOUSLY LIVING IN IRELAND SINCE THE LATE PLEISTOCENE, AROUND 30,000 TO 60,000 YEARS AGO

"

remained in the whole of Ireland, roaming the forests and uplands around Killarney. Thanks to strong protection measures and reintroduction programmes, the herds have recovered since then, and red deer can now be found, apart from their Killarney stronghold, in Connemara, County Donegal, County Wicklow and parts of Northern Ireland.

The Irish hare, having always been an inhabitant of open tundra and grassland, had no problem at all adapting to the growing peatlands, where it feeds on heathers, grasses and sedges. The Irish hare can only be found in Ireland and at the time of writing was still categorised as a subspecies of the mountain hare. Recent findings, however, have raised the question of whether the Irish hare deserves to be its own species. Fossil remains have shown that the Irish hare has been continuously living in Ireland since the late Pleistocene, around 30,000 to 60,000 years ago. This means the animal survived the various glaciation periods of the Ice Age and, unlike many other plants and animals, didn't migrate back into Ireland from neighbouring Britain or the continent after the last glaciation. During this long time, the Irish hare underwent a transformation from a permanent to a seasonal white coat; once snow and ice had become a rare

occurrence, it stopped the colour change almost completely. Only some animals have been observed to have a winter coat with a greyish tint, while the summer coat shows more of a russet brown. Today the Irish hare feels at home in a variety of habitats from the coast up into the mountains and roams fields and pastures as well as blanket bogs, raised bogs and heaths.

Birds

Two of the most likely birds to be encountered in peatlands are the skylark, singing its instantly recognisable song high up in the sky all summer long, and the snipe, which just like the skylark can mostly be heard and not seen. The summer visitors of this wading bird come from western Europe and Africa and make their presence heard by drumming, an eerie sound that echoes over the bog. It is created by vibrating feathers as the snipe performs rollercoaster flight patterns above its nest, which is well concealed in the high grass of tussocks in the wettest areas of blanket bogs and raised bogs. The snipes that spend the winter in Ireland, meanwhile, come from the Faroe Islands, Iceland and northern Scotland and only make themselves visible when they suddenly rise out of the heather, where they have been resting, and

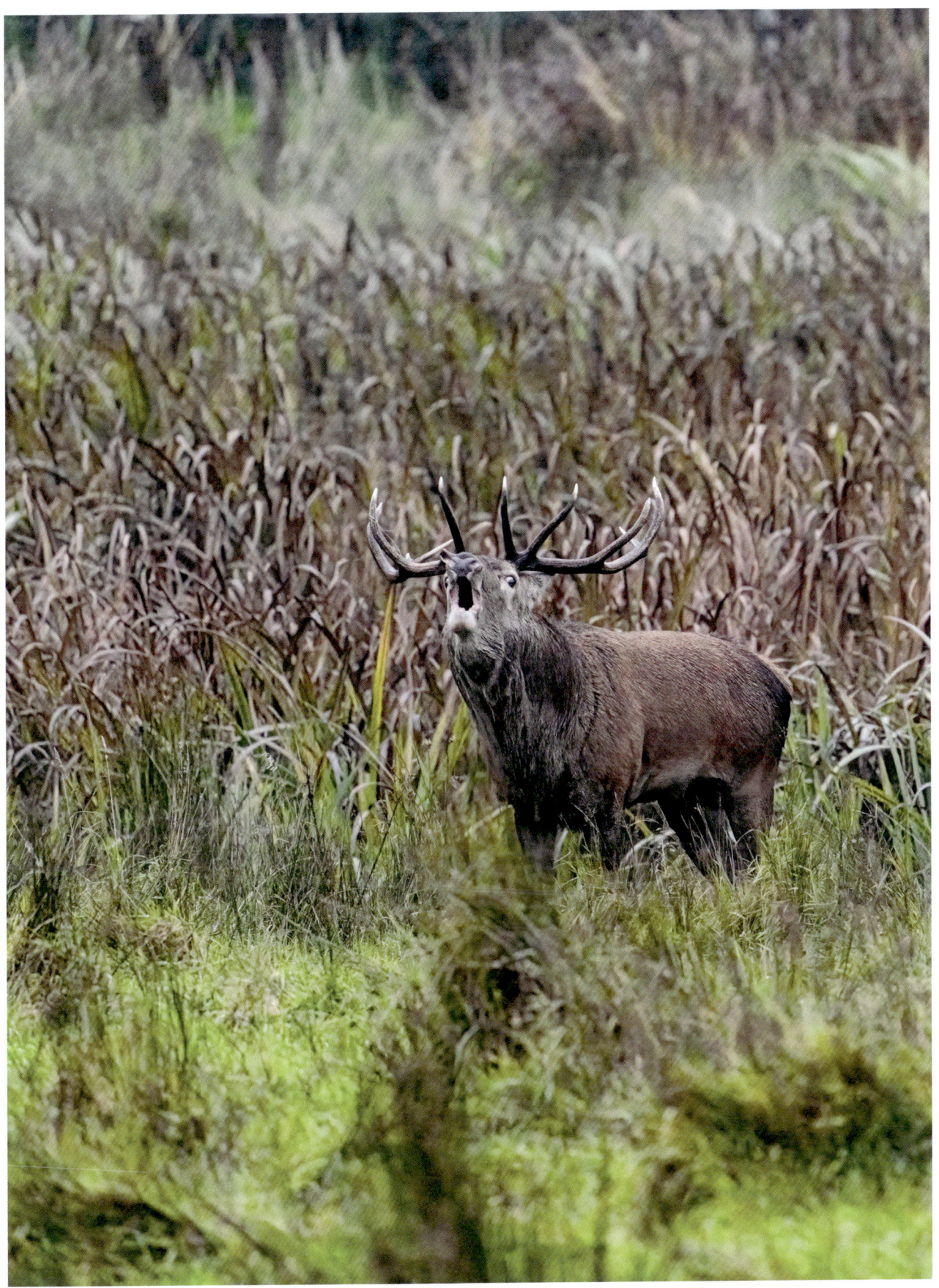

Red deer

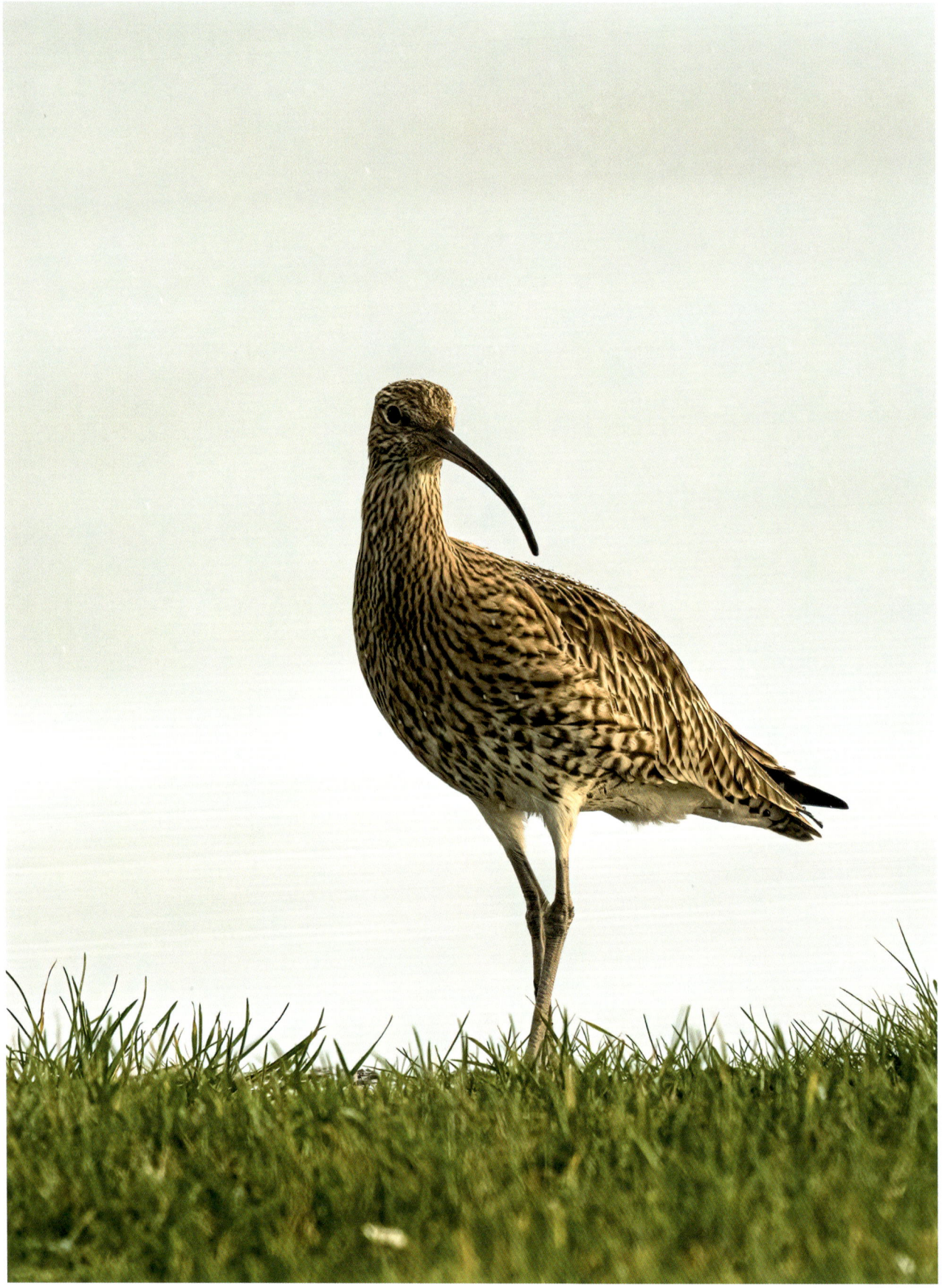

Curlew

"

THE RED GROUSE'S IRISH NAME, *CEARC FHRAOIGH*, TRANSLATES TO 'HEN OF HEATHER' AND VERY MUCH CAPTURES THE ESSENCE OF THE BIRD

"

make a quick escape in an erratic zig-zag pattern.

The curlew, once the quintessential peatland bird, has unfortunately almost disappeared from the landscape. The resident birds, whose numbers get a boost over the winter by visitors arriving from Scotland and Scandinavia, breed in wet meadows and peatlands, and their numbers have declined by 90% since the 1970s. Ever intensifying farming methods and large-scale afforestation programs in traditional curlew breeding grounds have been identified as the main reasons for this decline. In 2017, only 138 pairs remained, and researchers estimated that without intervention the curlew as a breeding bird would become extinct in Ireland by 2025. This realisation led to the foundation of the Curlew Conservation Programme, which over the past few years has had some success. While breeding pairs have further declined to just 100, the year 2023 saw a record forty-two chicks that reached fledgling stage – more than twice the number of the previous year. The story is similar for the lapwing. Just like the curlew, the lapwing is a ground-nesting wader that doesn't respond well to modern farming practices. The bird, with its characteristic headgear, was designated Ireland's national bird in 1990 but saw a steady decline in its breeding population that earned it a spot on the red list for endangered species. Breeding programmes, in collaboration with farmers, are starting to show some success: the population of lapwings on a site in County Wicklow has grown from just two breeding pairs in 2017 to seventy-seven nests in 2024, and the collaboration of the Ulster Wildlife Trust with a farmer resulted in the return of two lapwing pairs to a fen in County Down after the habitat had been restored over the past years.

Another typical but increasingly rare bird of the peatlands is the red grouse: a stocky bird, clothed in speckled brown, with a small bill and feathered legs. Its Irish name, *cearc fhraoigh*, translates to 'hen of heather' and very much captures the essence of the bird. Red grouse and common heather, also known as ling, are tightly intertwined. The birds feed on young heather shoots, flowers and seeds, and they build their nests in and from heather.

"

AROUND 8,000 GREENLAND WHITE-FRONTED GEESE – HALF THE WORLD'S POPULATION – SPEND WINTER NIGHTS ON THE SANDBANKS OF WEXFORD HARBOUR

"

Without heather there is no red grouse, and its disappearing habitat puts this elusive bird more and more under pressure.

The golden plover is another elusive bird but for different reasons. Ireland sits at the very southern edge of the bird's breeding range, and subsequently only a small number of golden plovers raise their chicks here; most of the population breeds further north in Iceland and on the Faroe Islands. Many of these birds, however – around 150,000 to 200,000 – come to Ireland's north-western counties to spend the winter.

Once known as the bog goose, the Greenland white-fronted goose was the emblem for winter in Ireland's peatlands. The birds breed in Greenland and travel to Ireland for the winter months, taking advantage of the comparably balmy temperatures and a rich food supply, namely the peatland staples white-beaked sedge and common cotton grass. Pressure from hunting and changes in peatland management, however, drove the birds from their traditional wintering sites in the northwest and midlands, and today few return to the bogs every year. Most of the visitors did find a new wintering home, though: the Wexford Wildfowl Reserve in the southeast of the country. There, around 8,000 Greenland white-fronted geese, which is about half of the world's population, spend their nights on the sandbanks of Wexford Harbour and in the mornings migrate the short distance to the fields of the North and South Slobs, where they feed on rye grass and the roots of buttercups and clover to restore their reserves for the long journey back to their northerly breeding grounds in spring.

In addition to those typical peatland birds, there are many other species that can be encountered on peatlands but which are equally at home in other habitats. Birds of prey like the hen harrier, merlin and buzzard regularly visit peatlands to hunt, and birds like mallard, teal, moorhen, coot and mute swan frequent ponds and lakes. Meadow pipit, reed bunting, sedge warbler, willow warbler and stonechat are not an unusual sight on bogs, and fens are home to a wide variety of songbirds including the latter and also wren, robin, linnet, dunnock, various finches, thrushes and tits as well as wood pigeon, collared dove, hooded crow, jackdaws and other visiting birds.

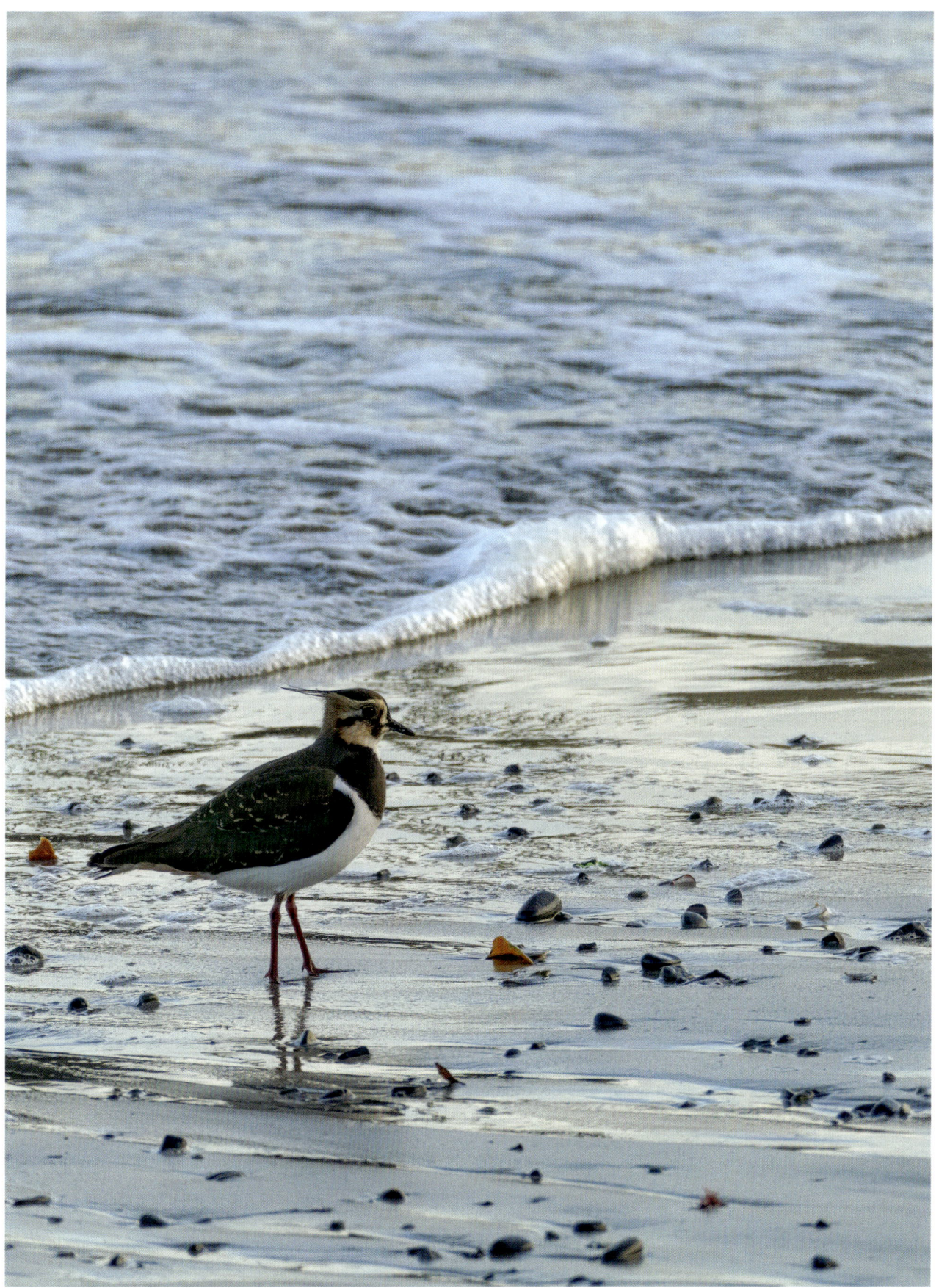

Lapwing

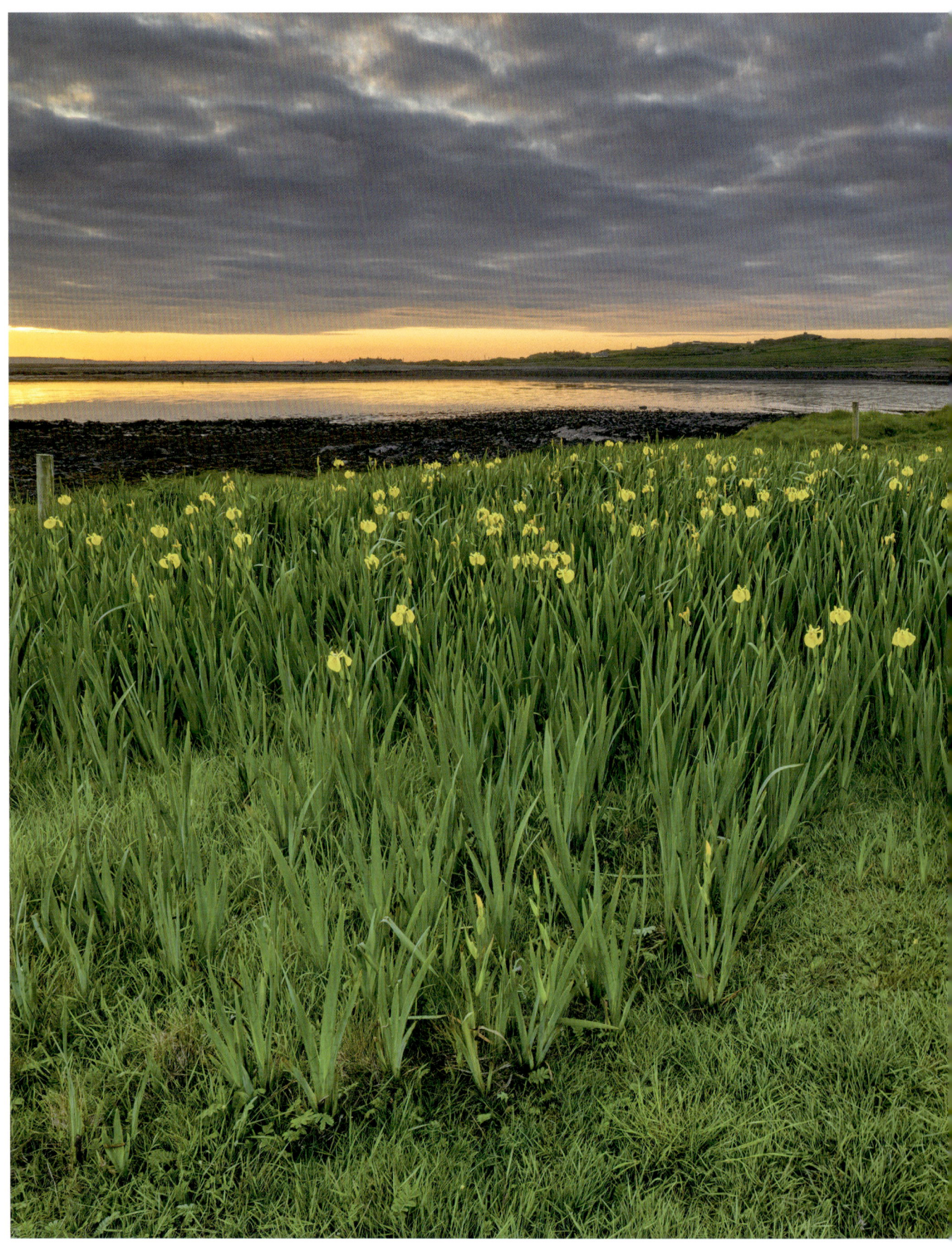

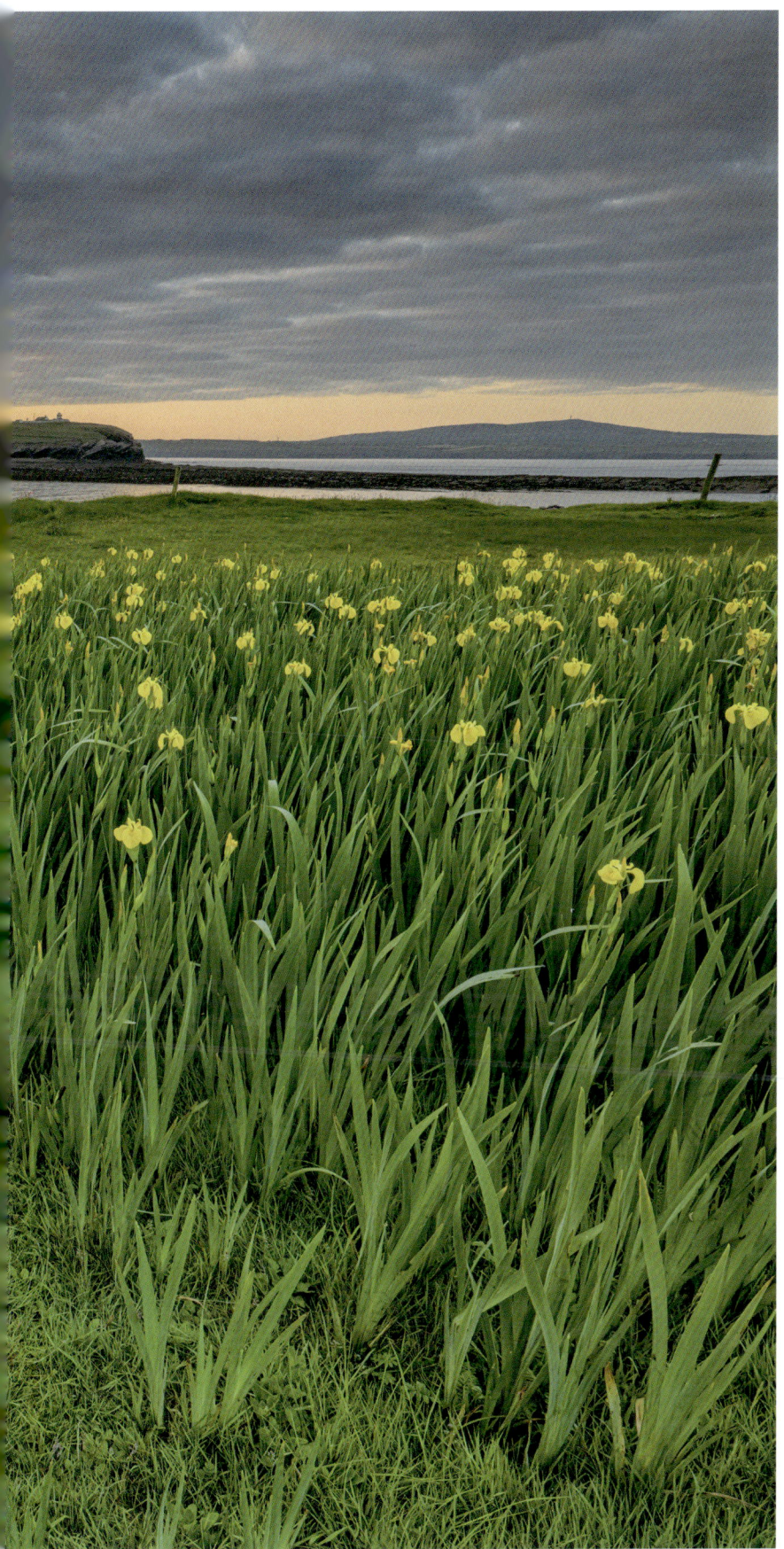

Flag iris

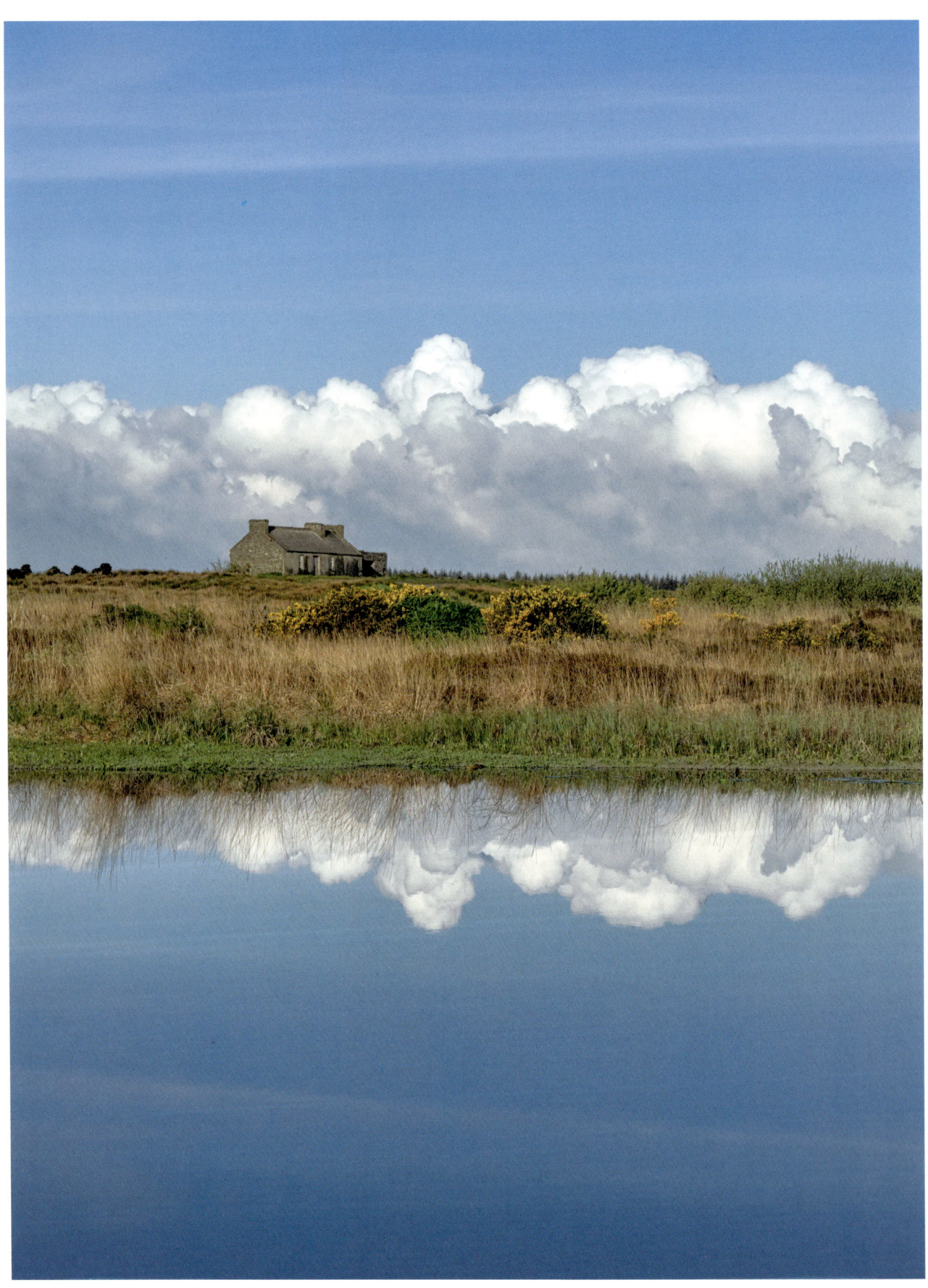

Old cottage in the bog, County Clare

Chapter Five
OF BOGS, MEN AND THE FUTURE

The relationship between men and the peat-lands is a complicated one, which shifted from a deep-rooted respect in pagan times to utter contempt with the arrival of the industrial age – men thought they were civilised, while the venerated landscape became a despised territory. This change of perception, however, didn't happen overnight and not in all minds. Especially in rural areas, residents mostly retained their respect for the land. In these places, life revolved around peat: it was the only source to heat the home and cook food and, in some areas, it was even a building material used to construct walls and roofs. Peat dictated the seasonal rhythm of the work that had to be done. Peat brought families and neighbours together. Peat made life possible.

The yearly cycle started in early spring, when the turf was cut and laid out in rows. After a few weeks, when the sods had formed a dry crust, the turf was 'footed' – arranged into little structures, each consisting of several sods – to dry completely over the summer. In August or September, depending how wet the weather had been, the dry turf briquettes were brought home and stacked safely for the colder months to come. This was – and in places, still is – a communal event. The residents of the village or townland came together to share the workload and make sure every family had enough fuel to make it through the year.

The turf harvest was labour intensive, back-breaking work, and extracting and drying enough peat for a whole year meant many days of hard labour had to be spent out on the bog. The *sleán*, a special kind of spade used to cut the turf sods, was eventually replaced by machinery, but this had only a small impact on the workload. Footing and turning the turf sods to make sure they dry from all sides, loading the dry briquettes on a trailer to bring the harvest home, and finally stacking the turf for storage is being done by hand to this day.

While this domestic use did cut scars into the surface of the bogs, the mostly small-scale removal of the peat gave nature the opportunity to pre-vail and the land a chance to heal. Turf was cut only once a year, and only the amount needed was removed. For the rest of the year, the bog belonged to the plants and animals, and once a turf bank was abandoned the flora recovered quickly and the bog came back to life. Having said that, turf

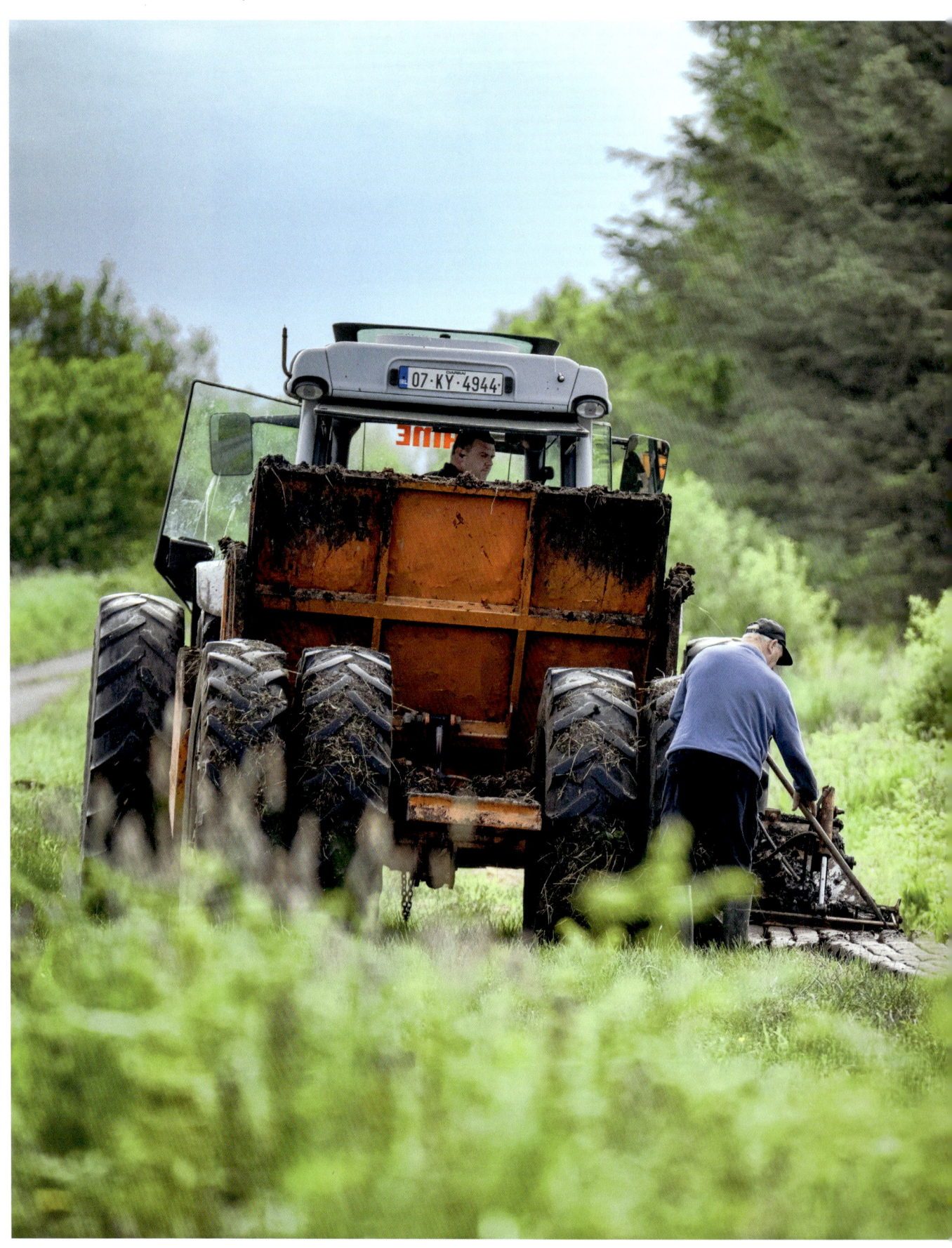

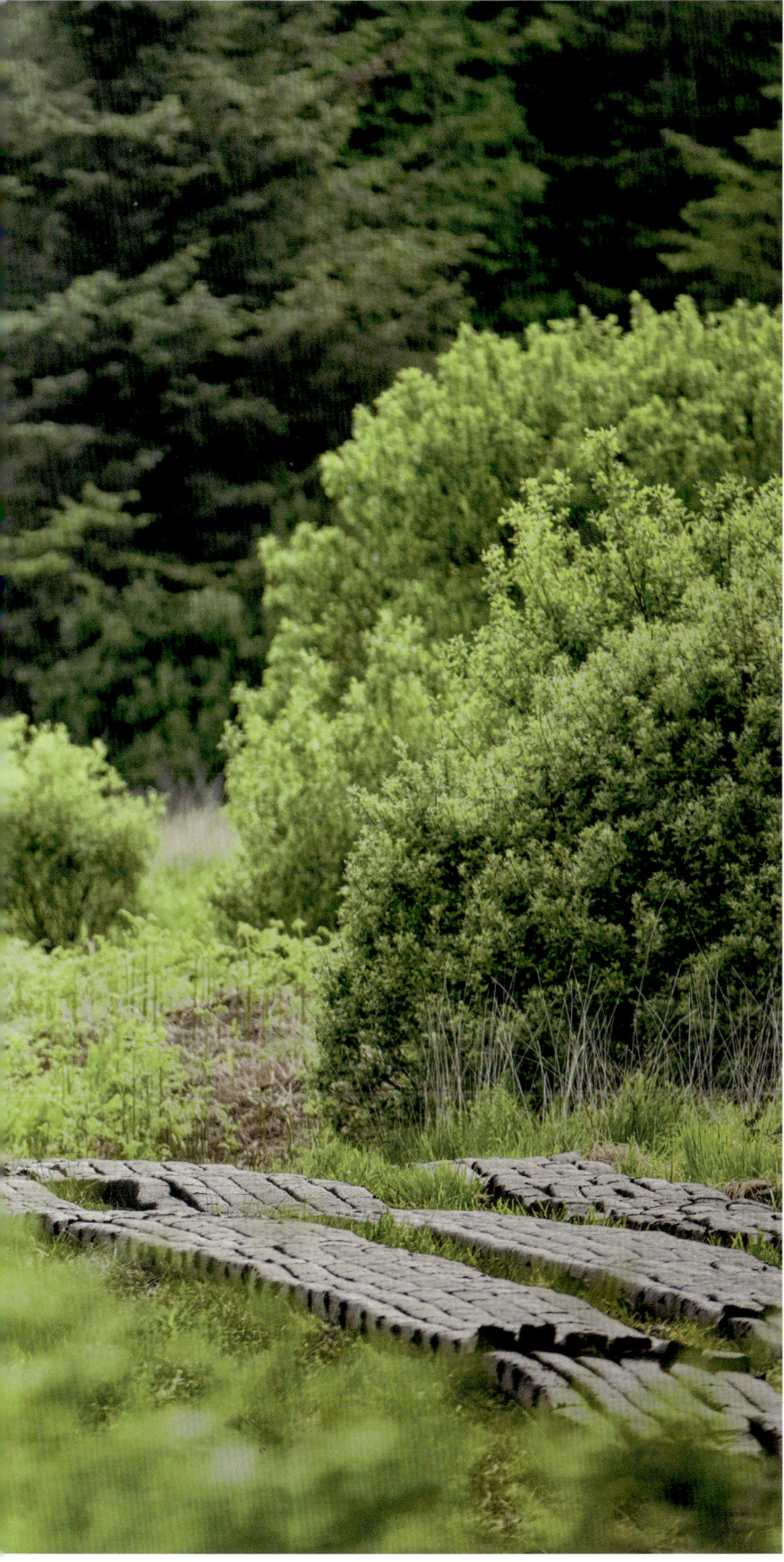

Peat harvesting with the
'sausage machine'

"

THE PRESERVATION OF OUR NATURAL HERITAGE CLASHES HEAD-ON WITH THE PRESERVATION OF ONE OF OUR LIVING TRADITIONS

"

harvesting for domestic use has become a fiercely debated topic over the past years. While any kind of turf cutting damages the bog, cutting one's own fuel from one's own piece of bog is a quintessential part of Irish heritage. From an environmental standpoint, it of course makes sense to stop any peat extraction; in reality, however, it is not that easy. It's a situation where the preservation of our natural heritage clashes head-on with the preservation of one of our living traditions. From what I have seen in my local bog, domestic peat cutting and a thriving peatland can exist side by side. In Shragh Bog, actively harvested turf banks intermingle with areas of actively growing bog with a thriving flora and fauna. With renewable energy options like domestic solar panels and heat pumps becoming more common, I believe that over time domestic peat cutting will disappear naturally. It will be a transition from one free energy source to another.

The idea of Irish peatlands as a symbol of poverty and backwardness arrived from the outside and mostly came from members of the ruling class and well-off members of the aristocracy. William King, a Church of Ireland archbishop of Dublin in the early 18th century, wrote: 'The smell and vapours … are very unwholesome; and the fogs … are commonly putrid and stinking … They corrupt our water …' Robert Dennis, author of a volume on industrial Ireland, described in 1887 'the stagnant, sodden, rotting substance which constitutes the bog-land …'

During the 19th century, this condescending attitude towards peatlands developed into a commercial way of thinking, and there was an increasing effort to make the perceived soggy wastelands into something useful. With improving drainage techniques, peatlands were transformed into farmland. Peat became more popular as a domestic fuel, which resulted in increased production of hand-cut briquettes, and eventually the first steps were taken to use peat in industrial settings. One such attempt

Opposite top left: Common cotton grass on cutover blanket bog
Opposite top right: Turf stacks
Opposite bottom left: Traditional turf cutting, Connemara
Opposite bottom right: Turf shed, County Roscommon

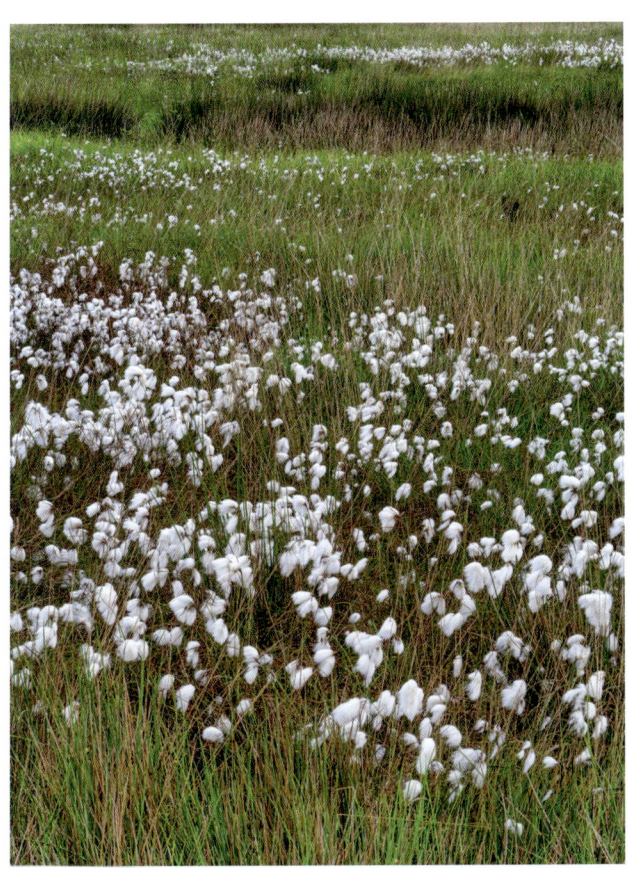

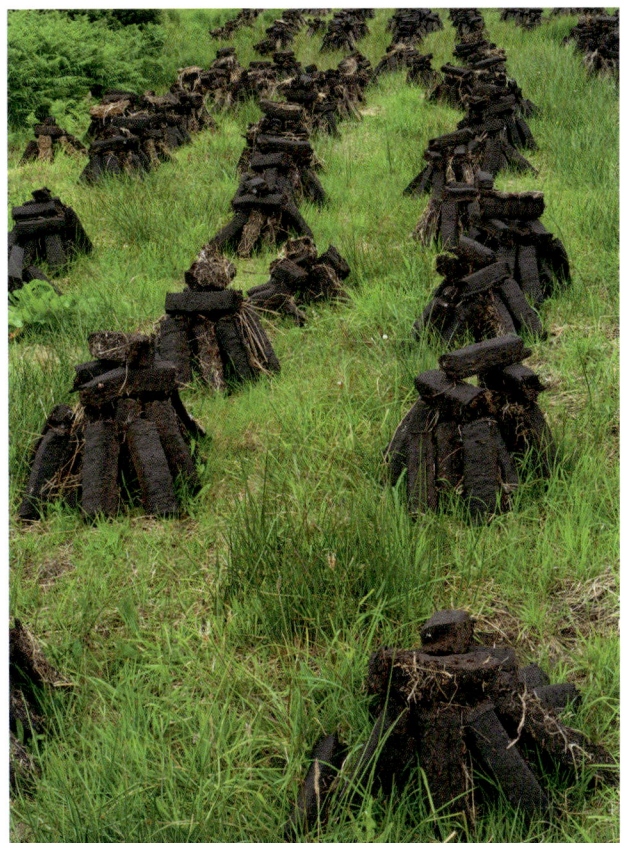

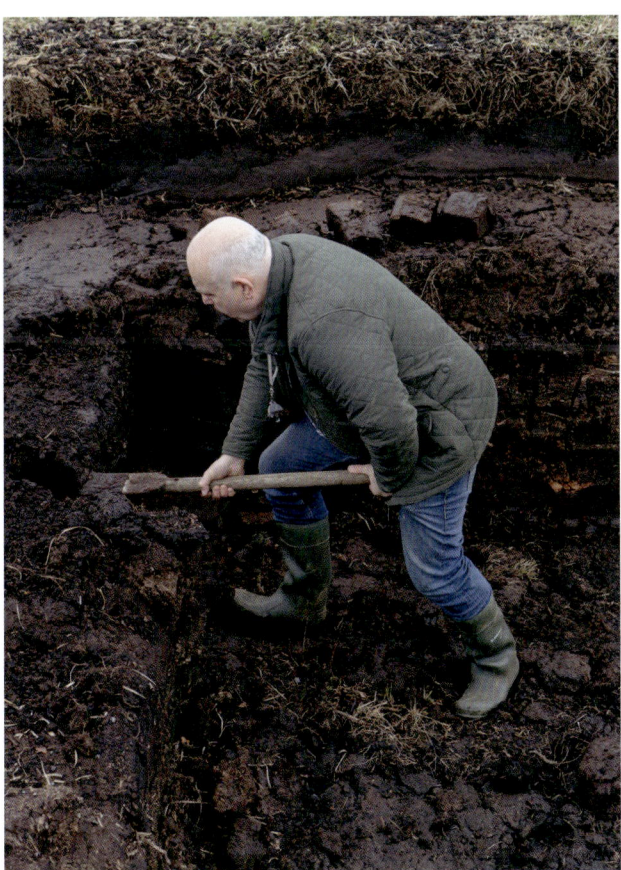

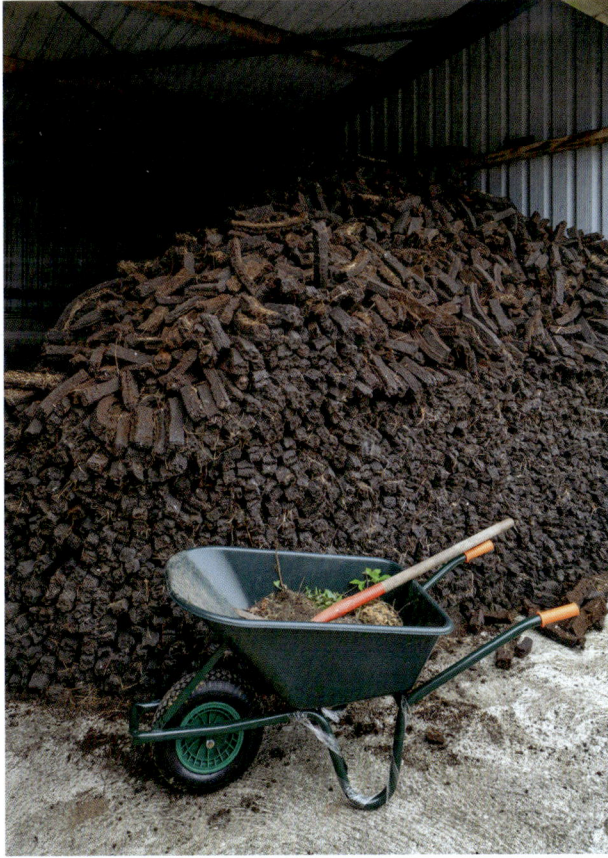

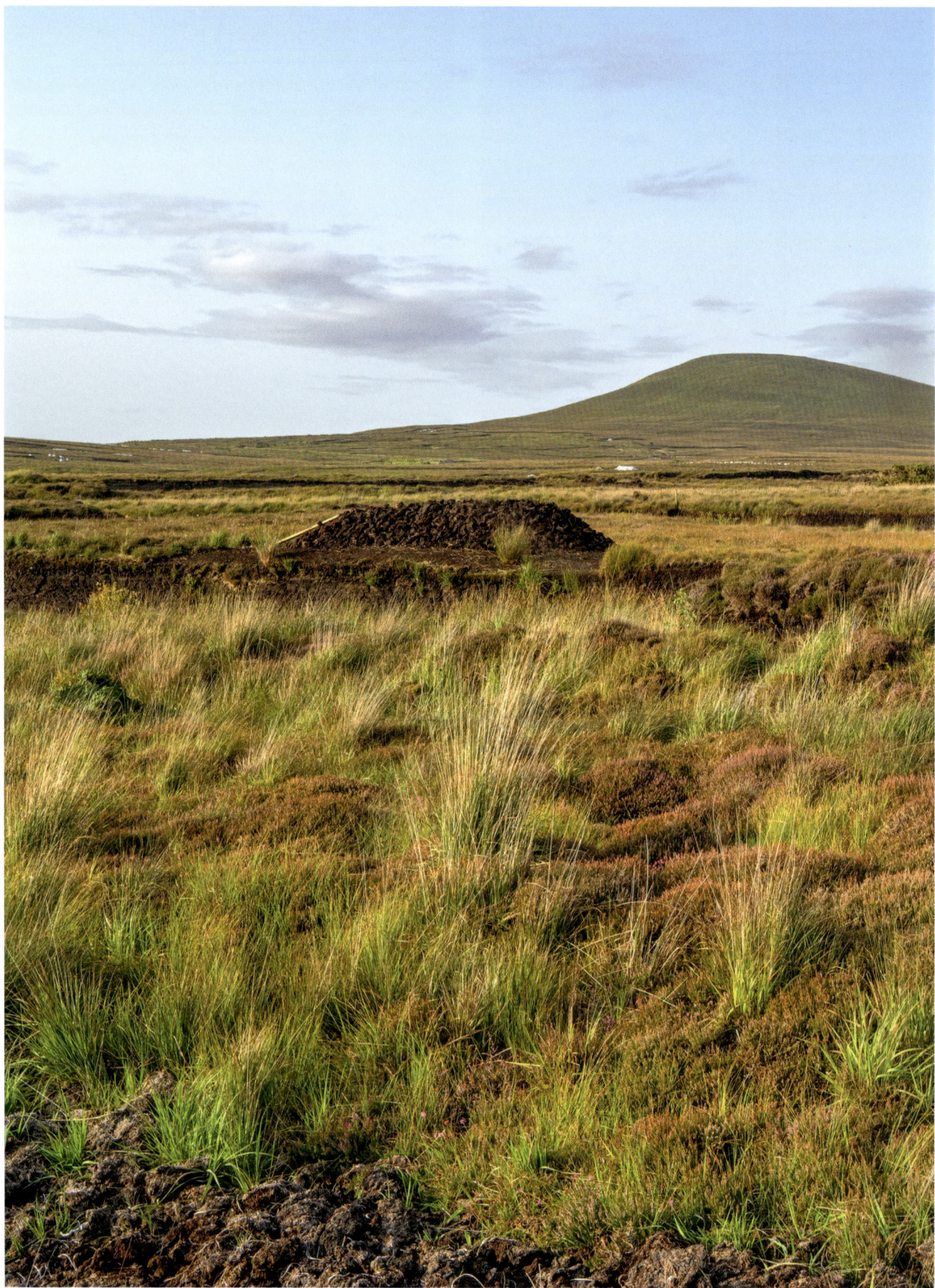

Turf bank, Achill Island, County Mayo

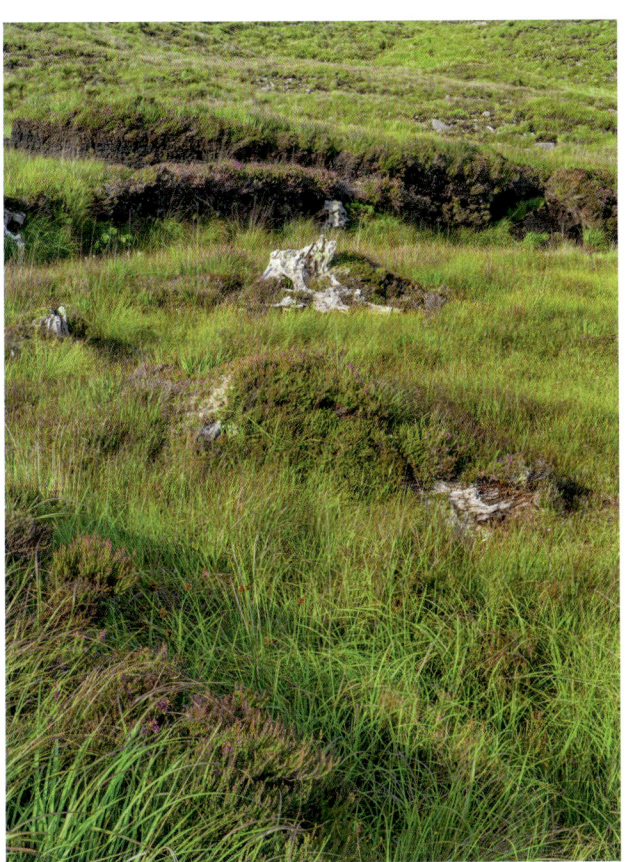

Abandoned turf banks with bog pine on Achill Island

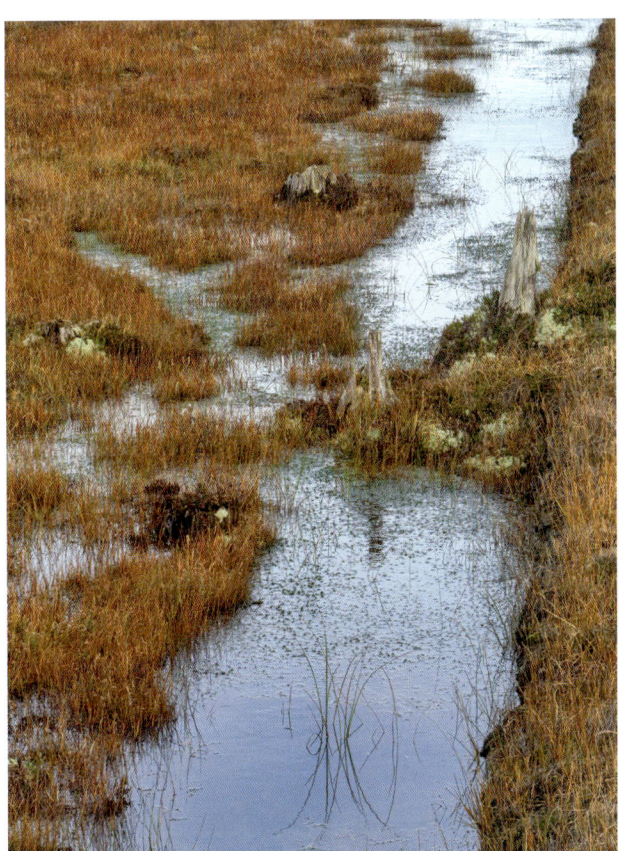

Drainage ditch in blanket bog, Tullaher Bog, County Clare

was made in Kilrush in County Clare, where peat was used in a steam engine to power the local mill.

After Ireland's independence in 1922, peat became a national strategy. The age of industrial peat harvesting and usage started with the formation of the Turf Development Board in 1934, which became Bord na Móna in 1946. These state bodies aimed to turn Ireland's peat bogs into commercially viable assets for the new free state and republic respectively. The outbreak of the second world war in 1939 and the resulting coal shortage further justified the government's plans, and increased efforts to make peat a pillar of the Irish economy were undertaken. To that end, dedicated Peatland Experimental Centres – one in Glenamoy in County Mayo and one in Lullymore in County Kildare – were put into place. These centres conducted a variety of research projects looking into how to improve and increase mechanical peat

extraction and diversify the usage of the extracted peat, and effective ways to grow crops and raise livestock on peatlands.

Implementing these ideas, however, didn't always go smoothly. It took over a decade to figure out and find the right tools for large-scale peat harvesting, in particular producing milled peat that could be shaped into standardised briquettes and used in power stations for electricity generation. Visits to Germany and Russia, two countries with considerable experience in industrial peat harvesting, helped to solve many of the hardware problems, and a white paper known as the First Development Programme, produced by the Irish government after the end of the Second World War, marked the proper start of the nationally orchestrated peat industry in Ireland. This paper included plans to develop twenty-four bogs for large-scale peat harvesting,

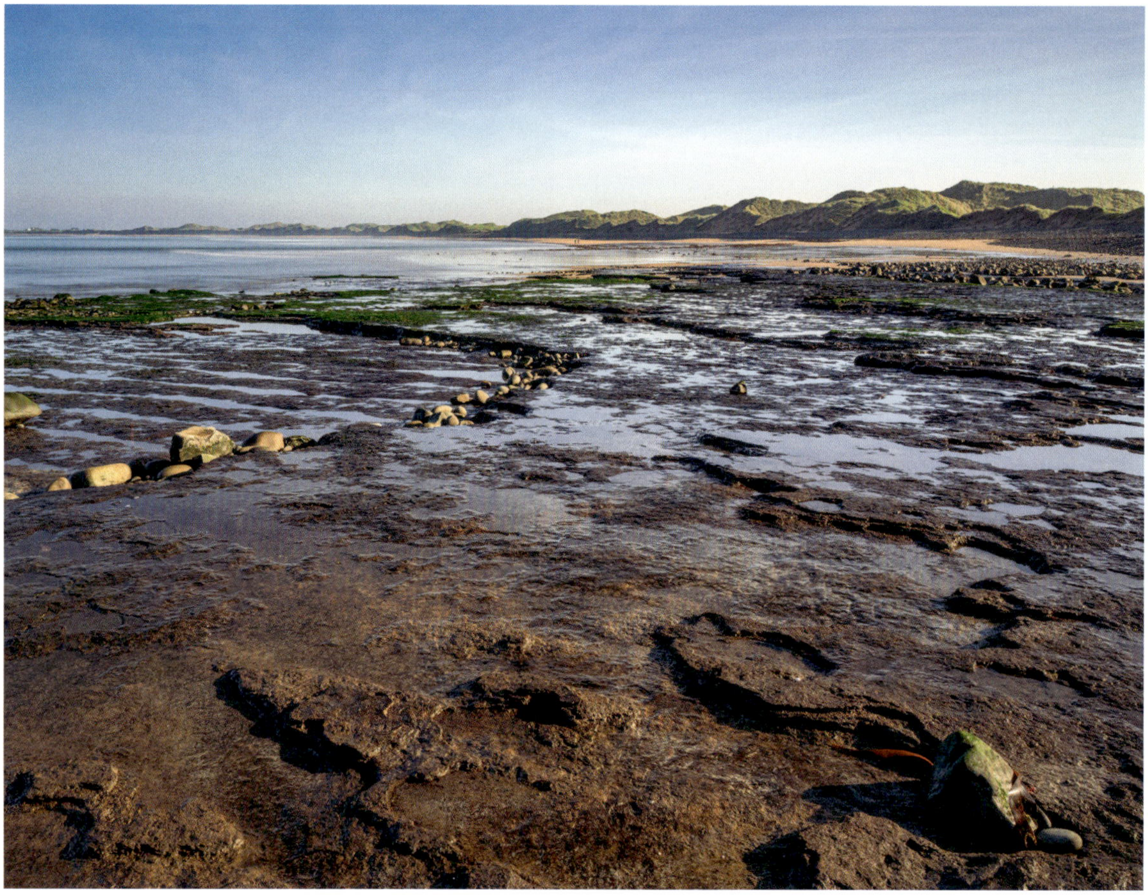

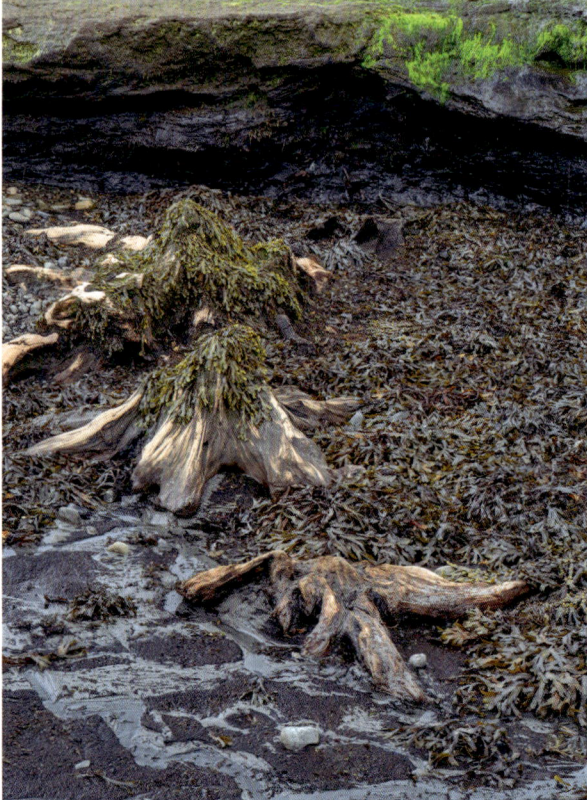

Above: Coastal peat, Doughmore, County Clare
Left: Eroding coastal peat on Loop Head Peninsula, County Clare

"

WHAT FOLLOWED WAS ONE OF IRELAND'S GREATEST ECONOMIC SUCCESS STORIES AND ONE OF ITS GREATEST ENVIRONMENTAL DISASTERS

"

with the goal to extract one million tons of peat every year. Also mentioned in the paper were intentions for the improvement of the Lullymore briquette factory, the building of a peat moss litter factory in Kilberry, County Kildare, and the planning and construction of the first peat-fired electricity stations at Portarlington, County Laois, and Ferbane, County Offaly. The document further included polices that all new electricity production should be based on turf, and that all housing, factories and institutions in receipt of state support, as well as local authorities, should be obliged to install turf-burning equipment.

What followed over the subsequent decades was one of Ireland's greatest economic success stories and one of its greatest environmental disasters. The vast raised bogs of the midlands were systematically drained and harvested to satisfy the growing need for electricity. At one time, ten peat-fired power stations supplied power to the midlands and parts of Ireland's east coast, each using up to one million tons of milled peat every year. In addition, harvested peat was used to produce a variety of peat products for gardening and home heating purposes. It is estimated that this industry decimated the raised bogs by 24% in only fifty years. Where peat harvesting was not, or was no longer, viable, the peatlands were transformed into farmland for livestock and crops.

In less than half a decade, the landscape of the midlands was utterly transformed. Today, a bird's

eye view reveals a checkerboard of green fields and immense plains of flat brown peat wastelands dissected by small roads and tracks of narrow-gauge railway lines, which had been built to transport the harvested peat to processing and power plants. The original extent of raised bog was once an estimated 308,742 hectares. Today only 1% of this area is considered to be intact, and only about 25% is deemed suitable for restoration.

The blanket bogs of the west and east fared only slightly better, which can mostly be attributed to their remote locations and inaccessibility. Originally almost 800,000 hectares of blanket bog covered the island of Ireland. A combination of turf-cutting, afforestation, farming and reclamation for other uses has left only 28% of these peatlands relatively intact. Traditional turbary rights (the right to harvest peat for domestic use) alone have claimed 36% of the blanket bogs, followed by large-scale Sitka spruce plantations, which have covered 28% of the original blanket bog areas. Intensified sheep farming and the development of wind farms have also degraded wide tracts of peatlands, especially in the uplands. While the harvesting of peat has been scaled back over recent years – a very controversial government decision, especially in rural areas – and no new Sitka spruce plantations are being planted, sheep farming and wind farm developments are still a threat to blanket bogs.

> **"**
>
> # THE FIRST AND MOST SUCCESSFUL OF THESE BOG RESTORATION PROJECTS IS THE ABBEYLEIX BOG PROJECT
>
> **"**

The latter decades of the 20th century saw the beginning of another change in attitude towards peatlands in Ireland. At the forefront of this journey to rediscover our respect and appreciation for the peatlands was the Irish Peatland Conservation Council (IPCC).

It all started in 1982 when a debate on the future of Ireland's bogs held at University College Dublin triggered the creation of the National Peatland Conservation Committee (NPCC). In the three years following the debate, the NPCC put together the first inventory of peatland sites in Ireland and an action plan for their conservation, and proposed a management structure for a campaigning organisation to implement the action plan. In 1985, the NPCC resigned and was replaced by the IPCC, a charitable organisation and company limited by guarantee that has become Ireland's voice for the peatlands. Over the past forty years, the IPCC has managed to designate 250,000 hectares of peatlands for conservation. It was a driving force in setting up a National Peatlands Council and preparing a National Peatlands Strategy, and has successfully embarked on a series of education programmes which have radically changed the public perception of peatlands.

The work of the IPCC has not only influenced national policy and public perception but also inspired a number of very real and tangible community-led projects. The first and most successful of these bog restoration projects is the Abbeyleix Bog Project. Abbeyleix is a town with just under 2,000 residents in County Laois, surrounded by a quintessential midlands landscape of fields, pastures and hedgerows. On the outskirts of the town sits Abbeyleix Bog, an approximately 200-hectare site of partially degraded and partially intact raised bog, bordered by semi-natural wet woodland.

The town and surrounding lands had been in the ownership of the de Vesci family, the local landlords, for some 300 years. Unlike many Irish landlords, the de Vescis looked after their tenants, and their first undertaking after acquiring the Abbeyleix Estate in the late 16th century was to move the entire town. Back then, Abbeyleix stood on the floodplain of the River Nore and experienced regular flooding events, which put the wellbeing and health of the residents in jeopardy. Under order of the landlord, the old town was abandoned, and the newly built town – the Abbeyleix of today – prospered. Among other

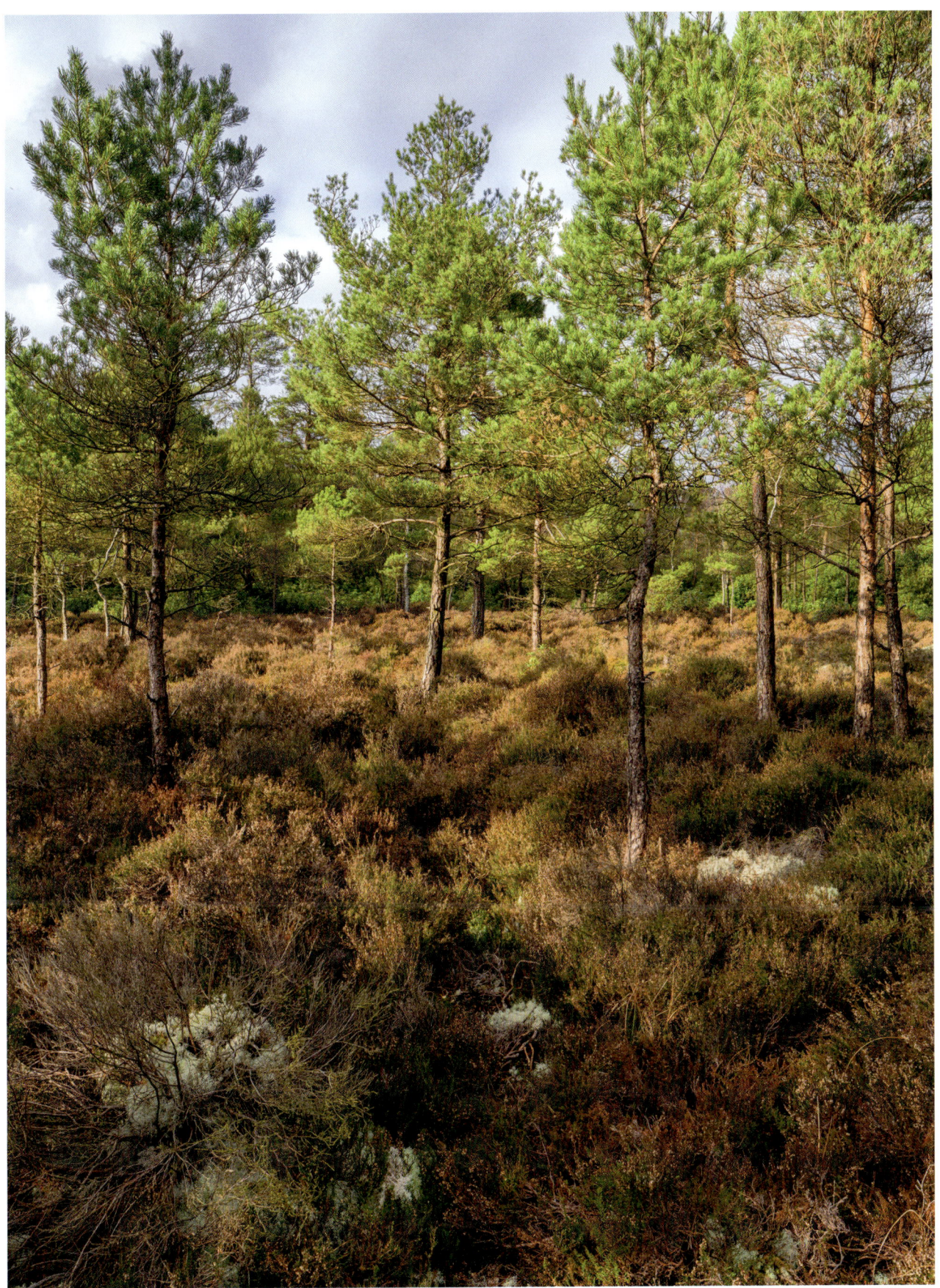

Abbeyleix Bog, County Laois

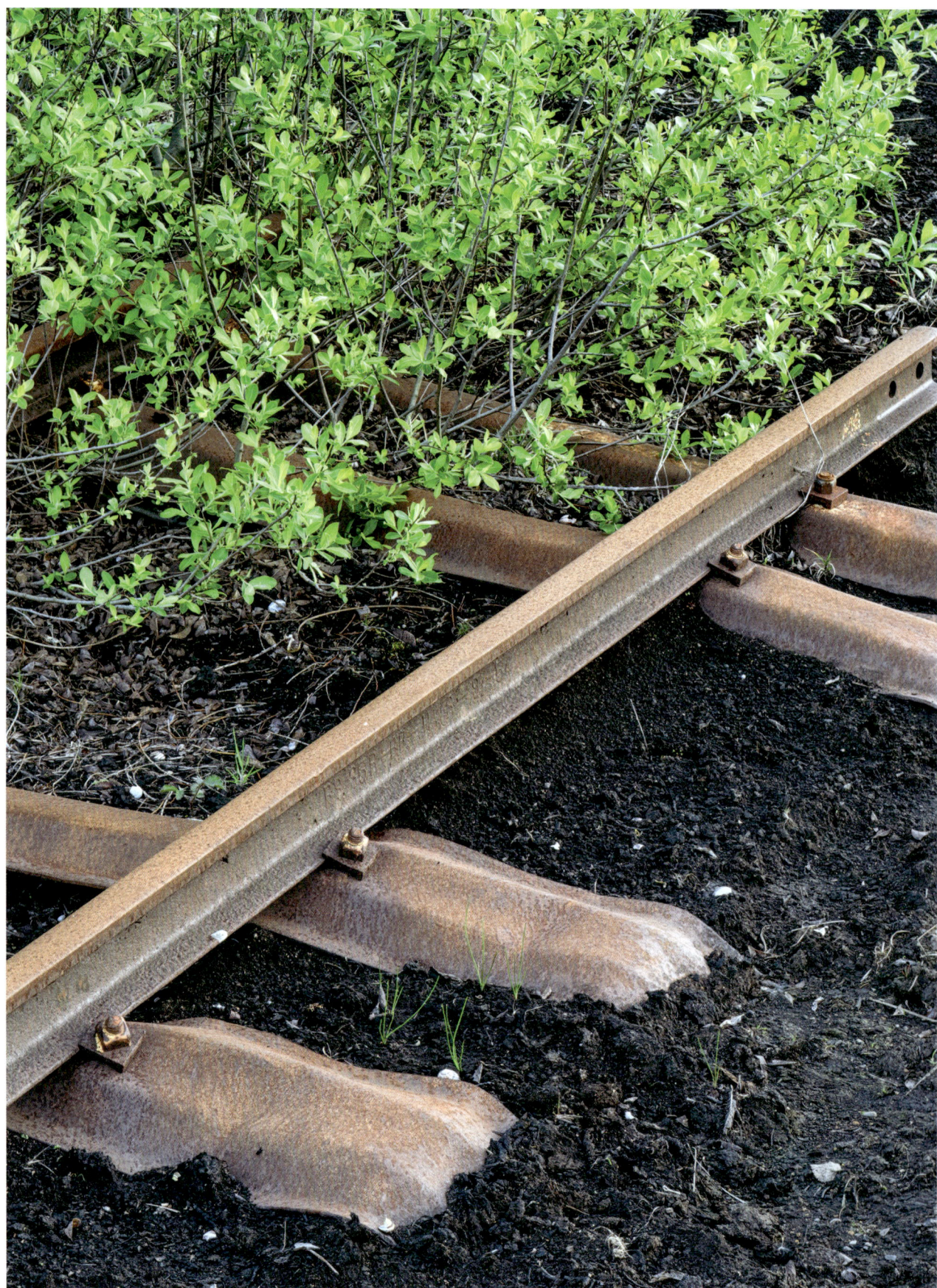

Abandoned narrow gauge railway, County Offaly

"

THE ABBEYLEIX RESIDENTS, JUST LIKE THEIR LANDLORD, VERY MUCH CARED ABOUT THE AREA IN GENERAL AND THE BOG IN PARTICULAR

"

improvements, the de Vescis connected the town to the railway network and developed various industries, including a carpet factory and a sawmill, ensuring employment and subsequently decent lives for the Abbeyleix residents. All this not only built a long-lasting, mutually respectful relationship between the de Vescis and the people of Abbeyleix but also resulted in a strong pride of place. The Abbeyleix residents, just like their landlord, very much cared about the area in general and the bog in particular. One Abbeyleix resident, interviewed for the Abbeyleix Bog Blog in 2016, summed it up: 'The bog has a lot of memories for the people of Abbeyleix.' Up to around the early 1970s, the bog was used for domestic turf cutting and as a playground for the children of the town: 'We played on it and had no concept that it could be dangerous … We painted our whole bodies black with wet peat and then ran out to the road.' More recently, Abbeyleix Bog was used by locals for walks and horse riding along the old railway line that runs in an elevated position through the bog. At the turn of the millennium, as climate change and biodiversity loss became ever more pressing topics, environmental awareness started to grow in the community, and many residents became conscious of the importance of their bog for carbon capture and preservation of biodiversity.

The story of the Abbeyleix Bog Project began to unfold in 1987 when the owner of the Abbeyleix Estate at the time, Tom de Vesci, came under growing pressure to sell his land to Bord na Móna

for industrial peat extraction to satisfy the growing demand for peat products in Ireland and abroad. Tom explained in an interview: 'I was approached many times by Bord na Móna to sell it after my father died in 1983, and I always refused. But eventually I was informed that Bord na Móna would be taking ownership via a compulsory purchase order at a somewhat lower level of compensation than I would get if I sold it "voluntarily" a few weeks earlier.'

With this sale, Tom de Vesci not only gave up his land, but he also gave up his and the community's dream to get the bog recognised as an important national site for biodiversity. One of Ireland's foremost environmentalists, geologist and botanist Professor John Feehan, had highlighted that Abbeyleix Bog featured large areas of intact raised bog very much worthy of preservation, and the National Parks and Wildlife Service had noted the bog as a potential area of scientific interest.

The new owners had no interest in assessing the option of preserving Abbeyleix Bog. In 1989, Bord na Móna started preparing the bog for future harvesting by cutting drains and installing a narrow-gauge railway to transport the harvested peat to a processing plant in nearby Portlaoise. For some residents of the town, the harvesting of Abbeyleix Bog was a prospect they could not accept. Around the time Bord na Móna started digging the first drains, a small informal group, which would eventually become the Killamuck Residents Association (KRA), formed to lobby for

"

THE WEEKS FOLLOWING THE STAND-OFF SAW THE COMMUNITY UNITING EVEN MORE AND ALSO REACHING OUT FOR SUPPORT

"

the preservation of the bog and its development as a local amenity. Over the following decade, they worked with the Abbeyleix Heritage Company and Laois County Council to develop a walking trail at the bog. They also managed to get members of the Irish and European parliaments interested in the matter, which brought the community's concerns to the attention of Bord na Móna. The following talks and negotiations dragged on and eventually reached a stalemate in 1999. As it turned out, this was the calm before the storm.

In 2000, the Abbeyleix Heritage Company tried to rekindle the negotiations on the walking trail and arranged a meeting with Bord na Móna. Little did they know that this meeting, which took place on 20 July, would mark the beginning of a chain of events of seismic proportions. Plans for the Abbeyleix Bog Trail were indeed discussed, but these quickly went up in smoke when on the following day, Friday 21 July, locals noticed two unfamiliar pieces of machinery on the bog, seemingly delivered by Bord na Móna overnight. The alarm was raised, and the community realised that the dreaded moment had finally arrived – Bord na Móna was preparing to start harvesting Abbeyleix Bog.

From there, events started to unfold quickly. On Sunday 23 July, a crane materialised on the only access road to the bog, blocking the entrance and impeding Bord na Móna from moving any more machinery onto the bog. It was claimed that the crane had unfortunately broken down during a locally organised bird survey session.

In an impromptu community meeting called on the same day, it was agreed that the residents of Abbeyleix would take action under the leadership of Killamuck Residents Association (KRA), and preparations were made to stage a protest and prevent Bord na Móna from accessing the site.

What followed was a stand-off between the people of Abbeyleix and Bord na Móna. On Monday morning, some fifty residents gathered on the access road to the bog. By lunchtime, the group had not only attracted more members of the community but also people from further afield as well as members of Laois County Council. An estimated 100 protesters prevented Bord na Móna from removing the crane and moving more machinery onto the bog. One protester remembers, 'It was an amazing thing and fantastic publicity for our cause.' A meeting between KRA and Bord na Móna representatives was arranged for Monday afternoon, during which a temporary ceasefire was agreed. The residents would remove the blockade, and Bord na Móna would stop moving machinery onto the site until both parties could agree on a long-term solution.

The weeks following the stand-off saw the community uniting even more and also reaching out for support. A community meeting to discuss the next steps was attended by almost 300 residents. To be able to properly engage with Bord na Móna, the Abbeyleix Residents for Environment Action (AREA) group was formed. A network was built consisting of organisations and individuals that supported the Abbeyleix community's cause, including the Environmental Protection Agency (EPA), An Taisce, the Irish Peatland Conservation Council (IPCC), Environmental Action Alliance Ireland,

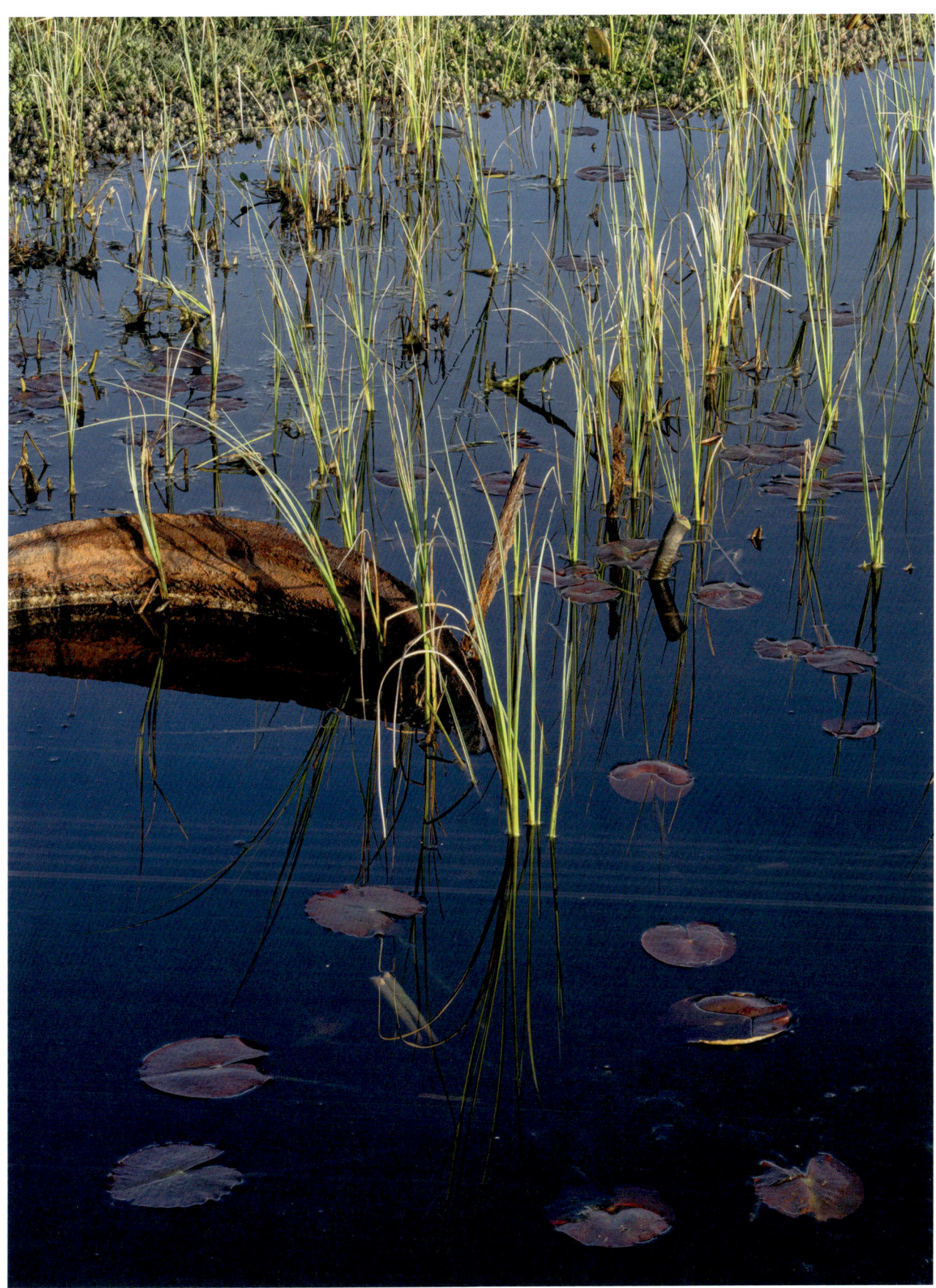

Bog pool as dumping ground

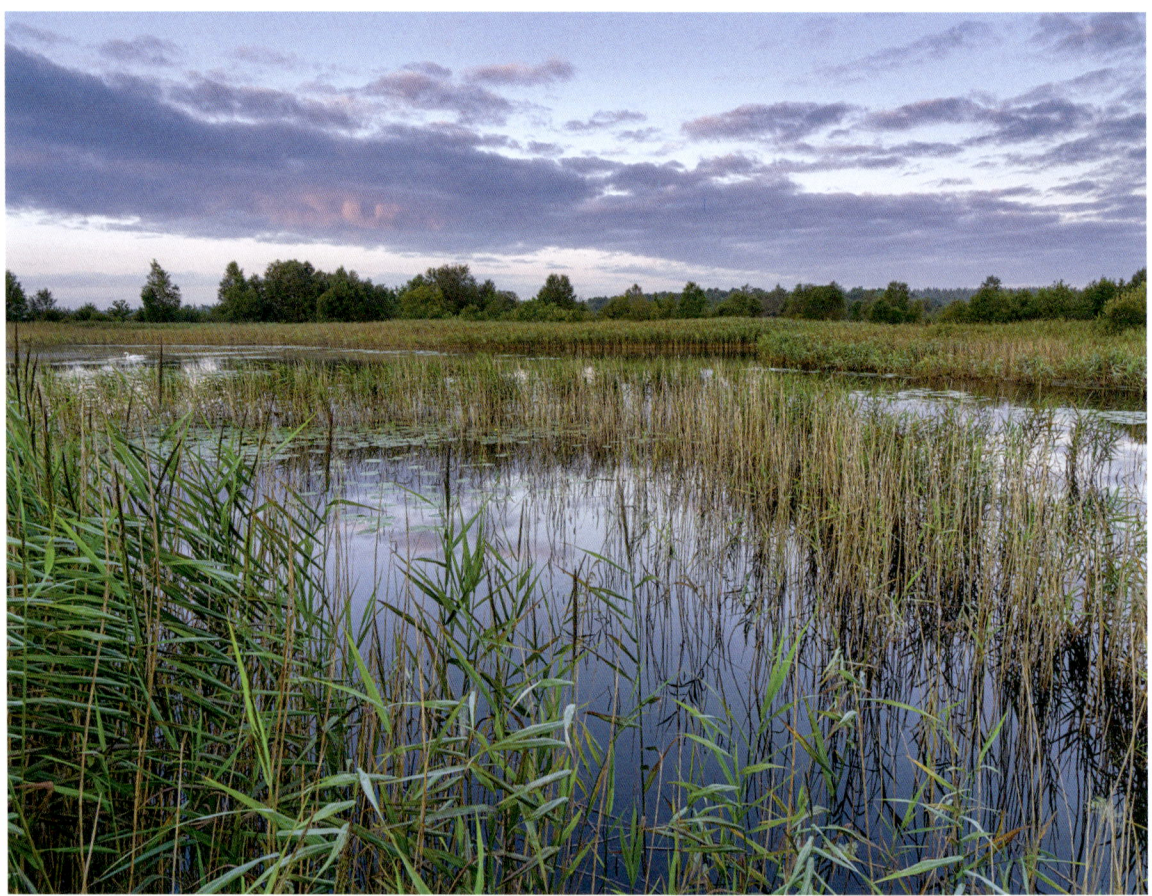

Lough Boora Parklands, County Offaly

the new owner of Abbeyleix Estate who had acquired the lands from Tom de Vesci in 1996, and independent ecologists.

The main goal of AREA was the preservation of Abbeyleix Bog, and their main demand was for Bord na Móna to prepare a proper environmental impact assessment. After several meetings, Bord na Móna eventually and reluctantly agreed to produce this document. The environmental impact assessment, compiled and published by Bord na Móna employees in 2001, stated that Abbeyleix Bog was of 'little or no conservation value'. AREA and its partners strongly disagreed and brought in two independent ecologists who, after assessing just 20% of the site, documented more than 500 species of plants and animals typical of intact raised bog and fen habitats, including a healthy population of the marsh fritillary butterfly, a protected species under EU law. In other words, AREA's environmental impact assessment stood in stark contrast to Bord na Móna's document and showed that Abbeyleix Bog had all the hallmarks of a habitat worthy of protection.

As a consequence, a complaint against Bord na Móna was successfully brought to the European Union (EU). The reasoned opinion from the EU, which was accepted and supported by the responsible department in the Irish government, led to Laois County Council seeking Bord na Móna to go through a proper planning process and apply for planning permission with An Bord Pleanála. This was a milestone event: for the first time in Irish history, a peat extraction development would have to go through the planning permission process.

In response to this decision, Bord na Móna took a high court action against both Laois County Council and An Bord Pleanála. The case was eventually heard in 2007; the resulting order, however, was never published. The idle time and the lack of a court order left all parties at an impasse, and neither AREA nor Bord na Móna were sure how to proceed.

Less than one year later, a series of fortunate events paved the way for what would become the Abbeyleix Bog Project. In August 2008, Minister for the Environment, Heritage and Local Government John Gormley visited Abbeyleix to open the new library, and AREA took this opportunity to present him with their plan to restore and retain the bog as a protected amenity and secure his support to reopen deliberations with Bord na Móna.

Coincidently, Bord na Móna was about to announce a change in its policy at the International Peatland Congress in Tullamore, County Offaly, in June 2008. This announcement would be the first step in the transformation of Bord na Móna from a fossil fuel to a climate change solution company. They would not open any new peat harvesting sites and would cease harvesting on existing sites on a phased basis. Indeed, by 2024 Bord na Móna had very much ceased all peat harvesting. The peat briquette factory in Derrinlough in County Offaly closed in 2022, and the last of the peat-fired power plants in Edenderry switched from using peat to burning biomass in 2023.

After the 2008 announcement, the conclusion of the Abbeyleix story happened quickly. With support from Minister Gormley, who also brought the National Parks and Wildlife Service's senior research scientist to join the effort, negotiations with Bord na Móna resumed, and, after a renewed study, Abbeyleix Bog was deemed worthy of restoration. The parties reached an agreement, and in April 2009, work began to block the drains in Abbeyleix Bog – almost exactly twenty years after they were cut – marking the first step in handing over the bog to the people of Abbeyleix. In December 2009, the Abbeyleix community signed a lease agreement with Bord na Móna that gave Abbeyleix Bog into the care of the community for fifty years, with the provision that the site was primarily used for habitat restoration. A public handover event took place three years later, in April 2012.

The Abbeyleix Bog Project CLG – consisting of the Abbeyleix Bog Management Committee, a Technical Advisory Group which includes two ecologists, local businesspeople and representatives from Bord na Móna, the National Parks and Wildlife Service, Laois County Council and the Irish Peatland Conservation Council, and a Board of Trustees – got to work and produced a Conservation Management Plan and Business Management Plan, completed in 2013, with the goal to maximise the ecological benefits to the local community in terms of education, employment, income and recreation. At the same time, various ecological and other baseline surveys were carried out to build a foundation on which changes could be measured.

What followed was a unique success story that made headlines not only in Ireland but also the rest of the world. A survey conducted in 2020 showed that, due to the careful restoration work, the area of active raised bog at Abbeyleix had grown from just over 1 hectare in 2009 to 13.78 hectares, an increase of 12.66 hectares in only eleven years. This made the Abbeyleix Bog Project a model for community-led conservation projects that is today attracting local, national and international visitors, nature lovers, students and researchers and has inspired similar projects in Ireland and abroad. Ironically, Bord na Móna would use the methodology implemented at Abbeyleix Bog for the company's own bog restoration programme.

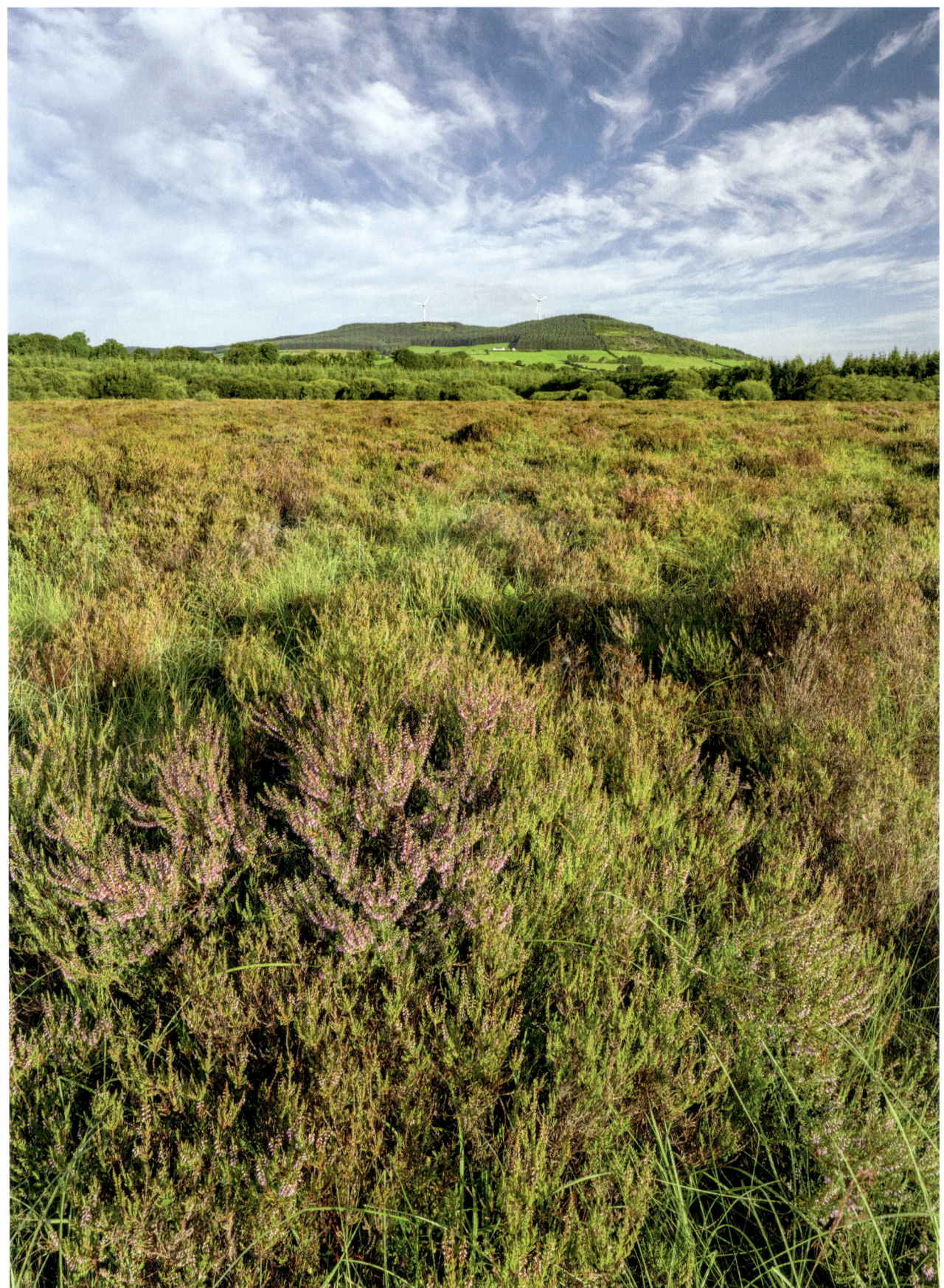

Raised bog in the valley and Sitka spruce plantations with wind turbines on the hills, Griston Bog, County Limerick

"

AN INTACT PEATLAND CAN CAPTURE AROUND 0.7 TONS OF CARBON PER HECTARE PER YEAR AND IF UNDISTURBED, IT CAN STORE IT FOR ETERNITY

"

The vision of the community for Abbeyleix Bog, born from a pride of place and local knowledge, has by now become national policy, and Bord na Móna and other semi-state bodies in Ireland have included peatland restoration in their strategic planning. The Living Bog Project, overseen by the Department of Housing, Local Government and Heritage, ran from 2016 to 2022 and aimed to restore twelve peatland sites in various corners of Ireland. Coillte, Ireland's semi-state forestry company, also runs projects with the goal to restore raised and blanket bogs that are currently covered by commercial spruce and pine forests. Coillte is also working in conjunction with the National Parks and Wildlife Service to create a wilderness area of over 10,000 hectares in the Nephin Beg range in County Mayo by replacing conifer plantations with native trees and restoring degraded blanket bog.

The main reason behind those restoration projects is, of course, the very real and pressing issue of climate change and the fact that peatlands are a most effective ecosystem for capturing carbon from the atmosphere and storing it for the long term. An intact peatland can capture around 0.7 tons of carbon per hectare per year and if undisturbed, it can store it for eternity. This means an intact peatland two metres in depth can store some

8,000 tons of carbon per hectare. In Ireland, it is estimated that peatlands hold 1,085 megatons of carbon, which corresponds to over half of all soil carbon on the island of Ireland locked away in only 16% of the land area.

While woodlands (which can store up to 300 tons of carbon per hectare) and the oceans (it is estimated that all the world's oceans store around 38,000 billion metric tons of carbon) are also important carbon sinks, both come with some setbacks. Woodlands release parts of the carbon they have captured back into the atmosphere over the summer months when their leaves fall to the ground and decompose, and their carbon capture rate very much depends on the age, size and growth rate of the trees. Trees we plant now only become noteworthy carbon sinks in a few decades.

Oceans store vast quantities of carbon, but there are some twists to this. Carbon in seawater gets permanently locked away only when it reaches the deeper layers of the ocean. In the surface layers, a constant exchange is happening between seawater and atmosphere: the colder waters around the poles absorb carbon, the warmer waters around the equator release carbon. Adding more and more carbon to the ocean also has considerable negative effects on marine life, especially shellfish and corals. This can set cascades in motion that could topple

"

PEATLANDS ALSO REGULATE LOCAL CLIMATE AND THE WATER BALANCE IN THE LANDSCAPE

"

complete marine ecosystems, with consequences that as yet elude our understanding.

In peatlands, on the other hand, the incomplete decomposition and the high-water table ensure that none of the captured carbon is released back into the atmosphere, and the stored carbon also doesn't have any negative impacts on the ecosystem. For this to work, however, peatlands must be intact, which first and foremost means they have to be waterlogged. The moment the water table goes down, the stored carbon reacts with the oxygen in the air and gets released into the atmosphere as carbon dioxide (CO_2). A study in 2009 showed that every year two billion tons of CO_2 escape from degraded peatlands worldwide – about 6% of global annual emissions.

Unfortunately, the climate benefits of peatlands are not entirely straightforward. While peatlands are an effective carbon sink, they also emit methane, the short-lived but potent greenhouse gas produced by the anaerobic decomposing of plant material that we met earlier as the will-o'-the-wisps. Methane is twenty-five times stronger a greenhouse gas than CO_2 and stays in the atmosphere for around a decade; the weaker CO_2, on the other hand, hangs around for thousands of years. However, compared to the burning of fossil fuels and farming, the amounts of methane emitted by peatlands are negligible and easily offset by the amount of carbon they can store.

In addition to being potent carbon sinks, peatlands have other benefits. They improve water quality by acting as a filter, removing excess nutrients and other pollutants as the water passes through the peat layers. Peatlands also regulate local climate and the water balance in the landscape. In times of high precipitation, they soak up and store large amounts of water, effectively preventing flooding; in times of little or no precipitation, they release the stored water back into their neighbouring habitats, preventing drought. Last but not least, the water evaporating from peatlands has a cooling effect, keeping the local climate more agreeable during times of high temperatures.

All of this, however, only works so long as the peat remains undamaged. Drained and compacted peatlands lose the pores in their structure and subsequently their water storage capacity – they become water repellent and so contribute to flooding events instead of mitigating them.

While most peatland restoration projects aim to preserve peatlands for carbon storage and biodiversity and are focused on bringing these habitats as close to their natural state as possible, some German and Dutch scientists suggest a more holistic approach that includes the revival of paludiculture. Having functioning ecosystems doesn't necessarily mean that humans need to retreat completely. The idea of combining nature restoration with land use might sound like an impossibility, looking at the

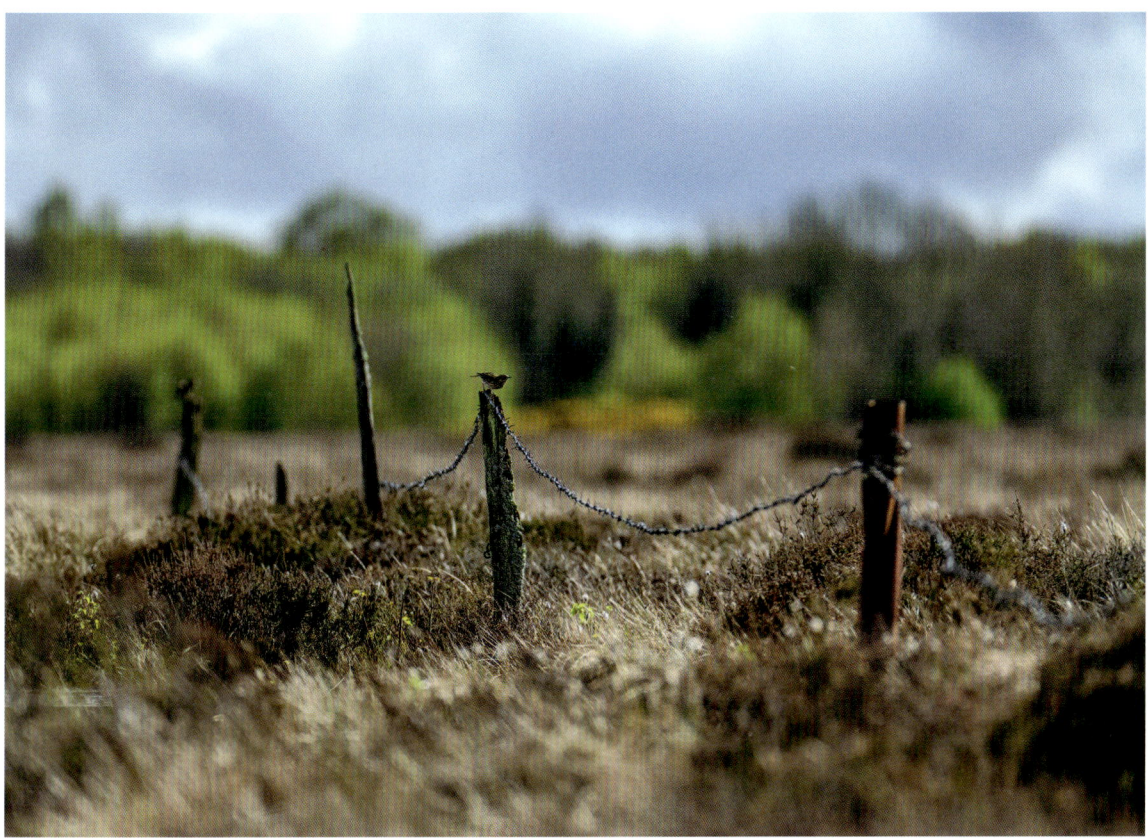

Meadow pipit at Lodge Bog, County Kildare

havoc modern agriculture is causing, but it has been done in the past. Preserving healthy wetlands as a haven for biodiversity and carbon capture, while also using these bogs, fens, wet grasslands and wet forests for their natural resources, could set an example that it is possible for humanity to thrive as a part of nature. And many of the raw materials found in peatlands could replace current less environmentally friendly components. Reeds and sphagnum mosses, in particular, have a variety of possible commercial uses ranging from medicinal and personal hygiene products to building and insulation materials. It is a vision of stepping back to a place in time when land use was practised in relative harmony and balance with the natural world. How this turns out – and how a balance between conserving and using peatlands can be struck – remains to be seen.

Ideas and visions like this raise another question: Why are we really doing what we are doing? Are we just looking for new ways to further exploit the natural world? Are we continuing to use peatlands in our usual short-sighted way, for our own benefits and needs, or have we learned something from the events of the past century that would suggest taking a less selfish, truly holistic and long-term view, and planning and acting accordingly? What is the right thing to do?

I like to believe – although there are many moments when I find it very hard to do so – that we can find a respectful and sustainable way to bring back a thriving natural world that provides all living things (including us) with what they need (but not necessarily what they want). A world to sustain body and soul; a place of magic and mystery.

AFTERWORD

There are two words I have used a few times in this book to describe peatlands: magic and mystery. They are not really my words – these and other similar terms have been used over and over, by many different people, to describe the natural world in general and peatlands in particular. Before modern science began to provide logical explanations for many things we see in the world around us, they must have been perceived as mysterious and magical. How does a caterpillar turn into a butterfly? How do you explain thunder and lightning? How can a plant create such beautiful flowers? How can birds fly? Why does the sky turn red before sunrise and after sunset?

These are just some of the questions our ancestors must have asked, and from these questions emerged what we call today pagan religions. While many of the explanations these ancient belief systems provided have been proven wrong by modern science, paganism had something we have lost along the way: a childish fascination with and respect for the world out there. Today, we have set ourselves apart; we feel superior to every other living thing on the planet, we believe we can control the world, and as a result we have thrown this very world out of balance.

The natural world exists in and manages to keep a delicate state of balance, ensuring the survival of a wide variety of very different species. When the deer population grows, so does the number of wolves to keep them in check; once the deer population has been decimated, the size of the wolf pack also shrinks. It is this game that repeats itself everywhere in nature and keeps things in balance for the greater good.

We have upset this balance by trying to tip it in our favour, and as we are now beginning to comprehend, this is not the way the world works. We are faced with a backlash – what we don't yet understand is how hard this backlash will eventually strike, and its consequences for us and every other living species.

The good news is that more and more people are trying to reconnect with nature and change to a more balanced lifestyle. Most governments worldwide have accepted anthropological climate change as a fact and know that the only solution to the problem is to stop, or at least drastically reduce, carbon emissions. Ever since scientists revealed that peatlands are formidable carbon sinks which could help us mitigate climate change, peatland restoration projects have been popping up everywhere. Research conducted in 2023 and published in 2024, however, showed signs that natural carbon sinks, including peatlands, are failing. They are failing because of the impact that climate change is having already. Scientists believe that widespread droughts and wildfires – 2023 and 2024 were the hottest years on record – prevented the ground globally from performing any notable carbon sequestration, and the world's oceans also seem to have reached the limit of carbon they can take on. The consequence of this discovery is that all climate change models need to be rewritten, and the outcome is unlikely to be in our favour.

Where do we go from here? I don't know, but I believe that the attitude our ancestors had towards the natural world, the very attitude first nations around the world still practise today, is the only way forward. We need to recover our mystery and magic, and with it a profound respect for the landscape and everything living in it.

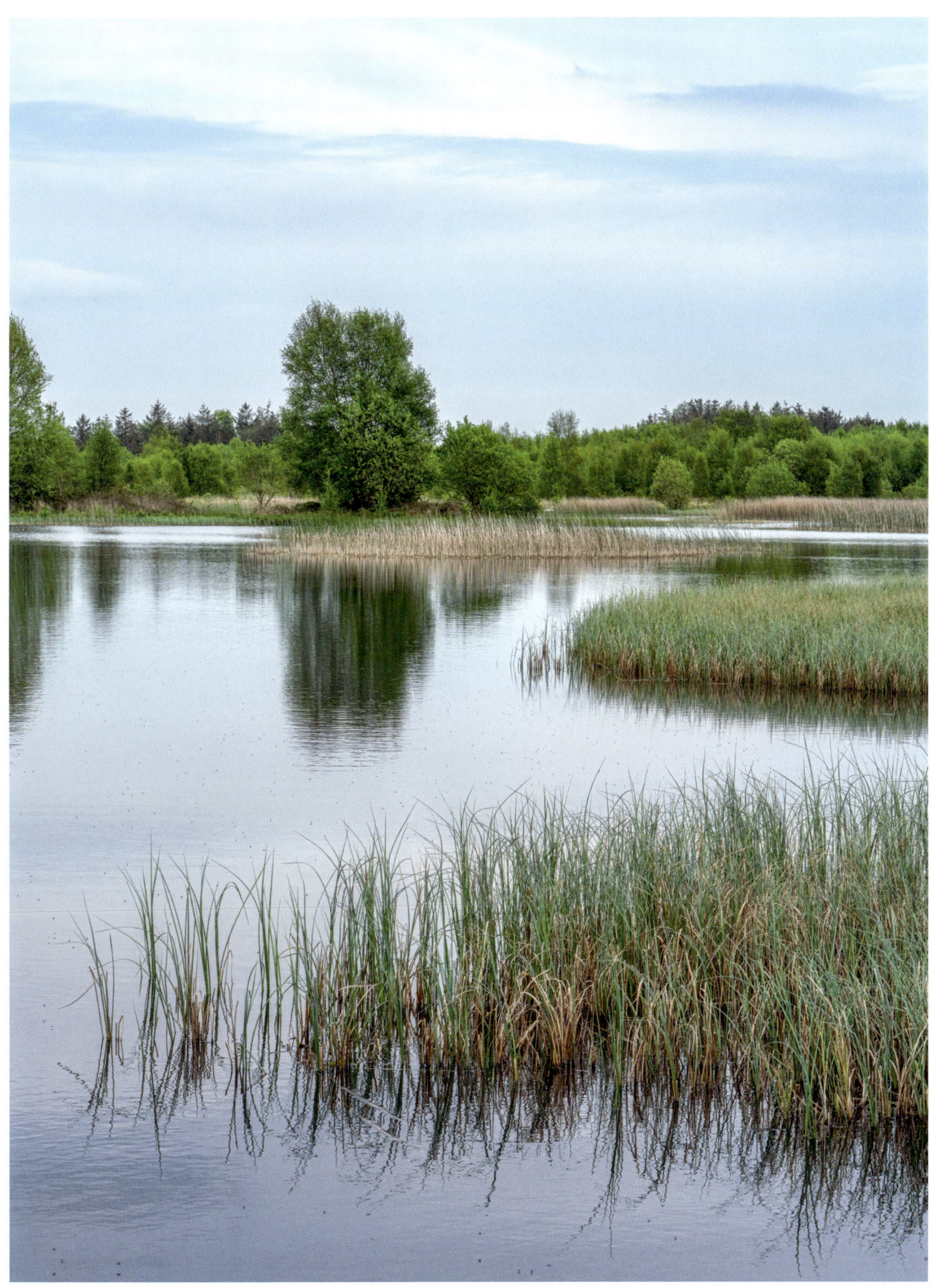

Tumduff Beag, Lough Boora Parklands, County Offaly

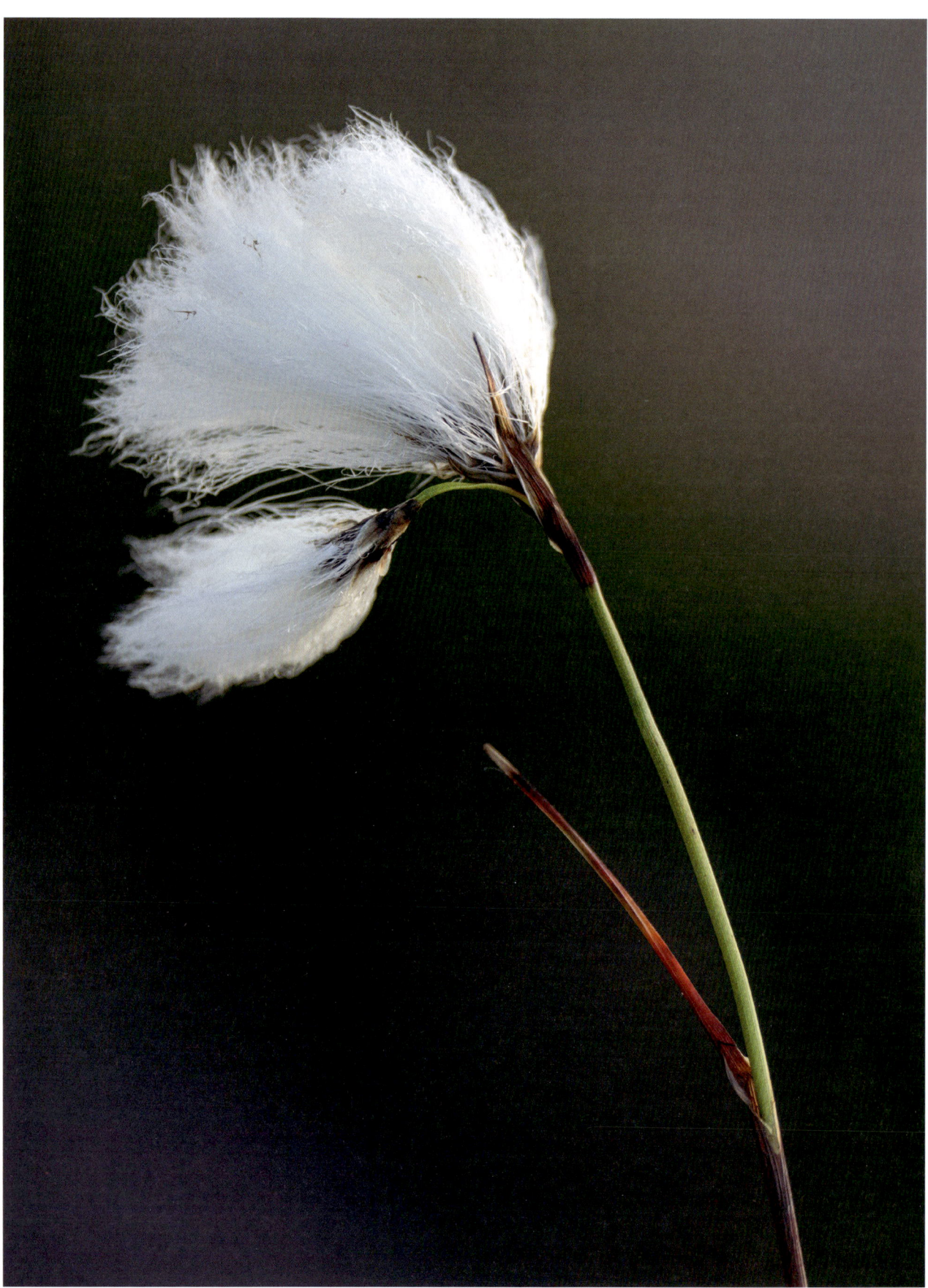

Common cotton grass

ACKNOWLEDGMENTS

First and foremost, I would like to thank Nuala Madigan and the team of the Irish Peatland Conservation Council for their support during this project. You made this book possible. Without you I would have missed quite a few photo opportunities, and your feedback on the manuscript was vital.

A big thank you also goes to Chris Uys of the Abbeyleix Bog Project for making the time to share the Abbeyleix story in every detail.

As always, a heartfelt thank you goes also to the team at The O'Brien Press and my agent Paul Feldstein for their help and support over the past twenty years (it really has been that long).

I would also like to send a thank you to Barbara Hurd (barbarahurd.com) for her inspiring writing and allowing me to use her words on the opening pages of this book.

Last but not least I'd also like to thank John Aston and the team at AstonECO for letting me pursue my book projects.

INDEX

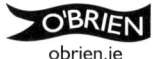
obrien.ie